Robert Bettmann

Somatic Ecology

Somatics, Nature, Humanity and the Human Body

VDM Verlag Dr. Müller

Impressum/Imprint (nur für Deutschland/ only for Germany)
Bibliografische Information der Deutschen Nationalbibliothek: Die Deutsche Nationalbibliothek verzeichnet diese Publikation in der Deutschen Nationalbibliografie; detaillierte bibliografische Daten sind im Internet über http://dnb.d-nb.de abrufbar.

Coverbild: www.purestockx.com

Verlag: VDM Verlag Dr. Müller Aktiengesellschaft & Co. KG
Dudweiler Landstr. 99, 66123 Saarbrücken, Deutschland
Telefon +49 681 9100-698, Telefax +49 681 9100-988, Email: info@vdm-verlag.de

Herstellung in Deutschland:
Schaltungsdienst Lange o.H.G., Berlin
Books on Demand GmbH, Norderstedt
Reha GmbH, Saarbrücken
Amazon Distribution GmbH, Leipzig
ISBN: 978-3-639-15025-4

Imprint (only for USA, GB)
Bibliographic information published by the Deutsche Nationalbibliothek: The Deutsche Nationalbibliothek lists this publication in the Deutsche Nationalbibliografie; detailed bibliographic data are available in the Internet at http://dnb.d-nb.de.

Cover image: www.purestockx.com

Publisher:
VDM Verlag Dr. Müller Aktiengesellschaft & Co. KG
Dudweiler Landstr. 99, 66123 Saarbrücken, Germany
Phone +49 681 9100-698, Fax +49 681 9100-988, Email: info@vdm-verlag.de

Printed in the U.S.A.
Printed in the U.K. by (see last page)
ISBN: 978-3-639-15025-4

Table of Contents

Preface

Every year the reporting on the environmental crisis gets more extreme. More and more species are going extinct. The polar ice is melting. Water is in short supply. The globe is warming. Despite increasing documentation of impending chaos and doom, there is reason to believe that humans could develop a healthier relationship to the natural world.

This book was written with the belief that how we think about things matters. To develop the best solutions to our environmental problems, we need to strive for the most complete explanations to understand the world, and our place in it. As an example: consider when you drive down the road and see an animal carcass. You can look at the body and see it as an accident. But dead animals are a regular output of our transportation system. They are not random accidents. They are statistically predictable accidents.

Albert Einstein wrote that people should, "Make things as simple as possible, but no simpler." It seems odd at first, but one can draw a straight line from our environmental problems, to the human attitude toward the rest of the natural world. And one can draw a straight line from the human attitude toward nature, to the human conception of the body. If one understands the importance of how we consider the body, we might be able to change the regular accidental outputs that together make up the environmental crisis.

In the development of Somatic Ecology I have been supported and challenged by teachers, colleagues, family, and friends. This text would not exist without Dr. David Orr (my advisor at Oberlin College) and Dr. Naima Prevots (my advisor at American University.) Thank you, also, to my family,

riends, and colleagues who have supported me during this materials long
gestation and development.

I recently read Lewis Mumford's <u>Sticks and Stones: a Study of American Architecture and Civilization</u>, and was charmed by the introduction he wrote for the 30th anniversary re-printing. Mumford admitted,

"Such a book could have been written only by a young man, with no reputation to risk, with no vested interest to protect, bold to the point of recklessness, and ready to intrude where professors, if not angels, would fear to tread. Doubtless if I had had a better sense of the difficulties or of my own limitations, I should have left the field alone."

In this first formulation of Somatic Ecology I recognize the integrity of the concept, and limitations in certain areas of the research and writing. I hope to address any mistakes or omissions in future work, and appreciate the reader's good will in bringing them to my attention.

Robert Bettmann April 16, 2009
Washington, D.C.

Introduction

The theory of Somatic Ecology is based on the hypothesis that if the body is valued as a knowledge source, a more ecologically balanced relationship may emerge between humanity and nature. The theoretical underpinnings of Somatic Ecology include earlier theories of environmental ethics, the study of the history of the body in western society, and the study of Somatic practices. The name 'Somatic Ecology' references the field of Somatics, and an earlier environmental theory (Deep Ecology) that inspired this investigation.

Significant background is necessary to establish the body as a motivator for an environmental theory, or a subject in discussing reversing environmental trends. The theories of Deep Ecology and Ecofeminism expose the importance of the human individual within environmental ethics. These theories, studied in concert with the history of how Western society arrived at the current relationship to the body, provide reason to consider Somatic methods to address the current environmental crisis.

Chapter 1 explores two major theories in environmental ethics, Deep Ecology and Ecofeminism, showing how the theories frame a revaluation of the relationship to the body. These theories will be exposed as a platform ripe for integration with current Somatic research.

Chapter 2 attempts to set a backdrop to modern developments, by establishing attitudes toward the body as documented in Jewish and Christian texts. The analysis includes texts from the Kabbalah for understanding of attitudes in the Judaic tradition, and texts from the Gnostic tradition and from he Apostle Paul to illuminate Christian views. These historical attitudes are addressed as parallels in the relationship between humanity and the body, and humanity and the non-human natural world.

Chapter 3 documents the simultaneous development of the scientific worldview with a widening divide from the body as a source of knowledge. Scientific and philosophical influences on the modern world are many. This chapter examines the scientific worldview as shaped by three seminal thinkers of the sixteenth through early eighteenth century: Galileo, Descartes, and Locke. Their work provides a window on changing perceptions and attitudes that distanced the body from Human understanding.

Chapter 4 presents the study of Somatics, beginning with an overview of the field, and considers three dance-related training modalities: Contact Improvisation, Skinner Release Technique, and Body-Mind Centering. These three training modalities are examined as both theoretical and practical methodologies for developing Somatic Knowledge. The focus in chapter 4 is on understanding the ideas about Knowledge generation these three movement philosophies have put forward, specifically, how they integrate mind and body to influence personal and community environments. The influence of Somatics has become increasingly significant beyond the dance field, in areas of movement education, body therapy, psychology, holistic medicine, and social work. There are older Somatic practices, and even newer ones, in diverse fields, that could be used to provide further documentation toward the ideas presented in this chapter.

In a prior version of this material there was a chapter that considered choreography and performance in relationship to the goal of validating and developing Somatic knowledge. That chapter needed such a thorough re-

6

working as to be detrimental to the whole were it included with these pages Performance practices exist that bring the ideas of Somatic Ecology to life in a visceral way. Appropriately directed research and presentation on that topic would meaningfully reinforce and deepen the arguments here.

Chapter 5 considers the theory of Somatic Ecology, exploring how placing value on the body can result in greater harmony within human and non-human ecologies. If the argument presented provides validity to the hypothesis, Somatic Ecology as a theory will have established the body as a critical site for investigation and intervention within the field of environmental ethics.

Deep Ecologist's argue that in order to address the environmental crisis, Western society must address the relationship of humans to the world. There is a simple logic to the notion that in order to change the relationship to nature, humans should start with the part of nature with which we are most intimate – our bodies. Somatic Ecology states that when we change the relationship toward our natural selves, broader societal change occurs. Such change is necessarily connected to a revaluation of the body as a source of Knowledge. Somatic Ecology combines insights from prior environmental theories regarding the roots of the environmental crisis with a new point of resolution – our selves. The investigation begins by considering current conceptualizations of the environmental crisis.

Chapter 1: Deep Ecology and Ecofeminism

This chapter analyzes two important philosophies that have emerged as driving influences in Environmental Ethics: Deep Ecology and Ecofeminism. These philosophies are analyzed in terms of their basic ideas, with a view toward increasing understanding about the impact of the existing human/nature duality. Ideas from each theory relevant to a consideration of the body as motivating subject are noted.

The science of Ecology is the study of interactions among biological systems. As such, Ecology is a study of the patterns of nature. Aside from its relevance within the scientific realm, Ecology has emerged as a motivator for Environmental philosophies. As David Pepper, the Environmental Historian put it: "Ecology conveys the universal principles about how humans ought to behave in nature." [1] More prescriptively the sociologist Ian Simmons explains that, "Ecology as a subject leads inevitably to certain values which can form the prescription for human behavior." [2]

Environmental Ethicists pose questions about the moral relationships between man and the natural world. As environmental theoretician Gregory Bateson noted, "the major problems in the world are the result of the difference between the way nature works and the way man thinks." [3] Those who work in the field seek to inspire or inhibit actions across the divide between the way that the biological and social worlds function.

---

[1] Pepper, David Modern Environmentalism pg. 240
[2] Simmons, Ian G. Interpreting Nature: Cultural Constructions of the Environment (New York: Routledge, 1993) 37
[3] Gregory Bateson interview at Lindisfarne, Long Island, 1976

The Norwegian philosopher Arne Naess coined the term Deep Ecology in 1973. In his seminal article "The Shallow and the Deep, Long-Range Ecology Movement: A Summary", published in Inquiry magazine, he outlined his ideas, and they have since developed and become mainstream within the Environmental Movement.[4] As a professor summarized the theory: the environmental crisis can be described as lots of puddles on the floor. Shallow Ecologists look at the puddles on the floor, and try to mop them up one by one. Deep Ecologists try to turn off the spigot that keeps dripping all over the floor.[5]

The Deep Ecology analysis includes a re-evaluation of critical components of modern life. Environmental philosopher Carolyn Merchant wrote, "Deep Ecologists call for a total transformation in science and worldview that will replace the mechanistic framework of domination with an ecological framework of interconnectedness and reciprocity."[6] Deep Ecologists believe that one must consider the overall character of human/non-human interaction if one is to capably address the environmental crisis. Deep Ecologists characterize human/non-human interaction as occurring within a paradigm of domination. Time will not be spent here considering whether the Deep Ecology assertion is justified, or necessary. Numerous existing texts handle that task more than adequately. This chapter begins accepting the assertion that the dominant relationship of man to nature results in the current environmental crisis.

Deep Ecologists believe that people need to change their relationship to the natural world. This will require wholesale changes in industry, commerce,

---

[4] Naess, Arne 'The Shallow and the Deep, Long-Range Ecology Movement.' Inquiry 16: (1973) 95-100
[5] Recollection by the author from class with Dr. David Orr, Oberlin College, Fall 1992.
[6] Merchant, Carolyn Radical Ecology (New York: Routledge, 1992) 11

politics, education, and technology. Deep Ecologists have identified three avenues consistent with the theory that lead toward 'turning off the spigot'. These are: the intrinsic value of nature, the norm of self-realization, and the need for a paradigm shift in humans' relationship to the natural world.

Deep Ecologists proceed from the belief that all of nature has intrinsic worth. They identify a utilitarian appreciation of the natural world as fundamentally unhealthy and systematically dangerous. Deep Ecologists argue that humans are obliged to respect the rights of nature.[7] They reject the dualistic view of Human and Nature as separate.[8] Succeeding chapters document that the reconnection of the Human within Nature is impossible without a revaluation of the body within the Human conception.

In the interview "Simple in Means, Rich in Ends", Naess investigates the pathway toward unification of the human with the natural, and describes it as "sympathy with other life forms."[9] The Deep Ecology pathway of sympathy with all life describes a non-utilitarian relationship to the world. In the Human sense of it, the Deep Ecologist's sympathy for life entails development of empathy for/with the natural world. This Deep Ecology pathway can be associated with Somatic training, in that both ask the rational consciousness to offer greater respect to the sensate being.

According to Naess, the "ultimate norm" of Deep Ecology is self-realization.[10] As another writer notes, "Deep Ecology presents environmental degradation as an abhorrent symptom of our alienation form the 'wild' parts

---

[7] Pepper, pg. 15
[8] Ibid. 17
[9] Naess, Arne and Sessions, George "Simple in Means, Rich in Ends" in Deep Ecology for the 21st Century Ed. by George Sessions (Boston: Shambhala, 1995) 28-9
[10] Devall, Bill Simple in Means, Rich in Ends: Practicing Deep Ecology (Salt Lake: Peregrine Smith, 1988) 23

of ourselves"[11] The pathway approaching self-realization, as defined by Deep Ecology, involves validating our animal self: our body.[12] Somatic practices result in re-connecting the human to its own body, developing inward and outward sensitivity. Through the empathy we feel as a live being we find sympathy with all life forms. The norm of self-realization clearly connects – through the body - to the norm of sympathy with all life. The Deep Ecology norm of self-realization requires a re-invigoration of one's animal self. Somatics and Deep Ecology share recognition of the body as the necessary site for focus.

Deep Ecology is not only an internal, and possibly passive, process. Deep Ecology theory starts with the person, framing the necessity of change within the paradigm of how humans relate to the natural world. According to Deep Ecologists, however, the pathways of sympathy and self-realization must lead to real change in how people live in the world. The parallel norms of Somatics and Ecology are clear in their human re-valuations; by extension, and of import to Somatic Ecology, Somatic training can lead to a change in how people live in the world.

Integrating theory into specific practice is one of the biggest challenges for Deep Ecology.[13] To augment the principles of Deep Ecology, outlined by Naess, Bill Devall and George Sessions created a Deep Ecology platform.

---

[11] Lahar, Stephanie "Roots: Rejoining Natural and Social History" in Ecofeminism: Women, Animals Nature ed. Gaard, Greta (Philadelphia: Temple University Press, 1993) 109
[12] The sense of the human/nature divide as a root of the crisis is widespread within the environmental community. As reported in the Earth First! Journal (Spring 1989), the founder of the environmental activist/terrorist organization Earth First!, David Foreman succinctly declaimed to his followers: "You're an Animal – Be Proud of It!"
[13] Devall, Bill Simple in Means, Rich in Ends: Practicing Deep Ecology (Salt Lake: Peregrine Smith, 1988) 27

The Deep Ecology platform provides guidelines for daily practice.[14] The platform has eight planks, including a basic statement valuing all life, and a call to arms for those who believe in the platform. As published in Zimmerman's Contesting Earth's Future, the planks of the Deep Ecology platform are[15]:

1. The well-being and flourishing of human and non-human life have value in themselves.

2. Richness and diversity of life forms contribute to the realization of these values and are also values in themselves.

3. Humans have no right to reduce this richness and diversity except to satisfy vital needs.

4. The flourishing of non-human life requires a smaller human population.

5. Present human interference with the non-human world is excessive and getting worse.

6. Policies must be changed. Economic, Technologic and Ideologic structures must change.

---

[14] The Deep Ecology stand reminds one of the famous Lincoln quote: "I don't care for a man's religion if it doesn't make a difference to how he treats his dog."
[15] In Zimmerman, Michael Contesting Earth's Future: Radical Ecology and Postmodernity. (San Fransisco: University of California Press, 1994) Pg. 24-5

7. The ideological change will be mainly that of appreciating life quality over standard of living.

8. Those who subscribe to the foregoing points have an obligation to try to implement the necessary changes.

Cultural critic Theodore Roszak eloquently argued that "most of our master ideas about nature and human nature, logic and value become so nearly subliminal that we rarely reflect upon them as human inventions, artifacts of the mind." [16] Deep Ecologists ask the individual to re-evaluate deeply ingrained beliefs and values about humanity, and the role of the individual. The Deep Ecology philosophy, and platform, asks Humans to reconsider our core values, including our definition of 'quality of life' [17].

Deep Ecologists ask humans to question, challenge, re-evaluate and update longstanding beliefs and practices. As Michael Zimmerman wrote, "Deep Ecology, then, seeks to overturn the major Western categories that are apparently responsible for humanity's destruction of the biosphere: anthropocentrism, dualism, atomism, hierarchalism, rigid autonomy, and abstract rationalism." [18] Deep Ecologists believe that a richer life develops as one becomes attuned to the fuller life that envelops people living on the earth. The changes that must be made, the things one must give up, the

---

[16] Roszak, Theodore The Cult of Information (New York: Pantheon Books, 1986) 106
[17] Such questioning is not unique to the Environmental Movement. Robert Kennedy's speech at the University of Kansas on March 18th, 1968 argued that, "...[the] Gross National Product measures everything except that which makes life worthwhile. And it can tell us everything about America, except why we are proud that we are Americans." Audio recording of speech published in Robert F Kennedy In His Own Words, CD (Jerden 1995) Track 5
[18] Zimmerman, Michael A. "Deep Ecology and Ecofeminism: The Emerging Dialogue" in Reweaving the World: The Emergence of Ecofeminism Irene Diamond and Gloria Feman Orenstein Eds. (San Fransisco: Sierra Club Books, 1990) 141

compromises, are willfully made in response to the riches that are gained with the spiritual and emotional growth gained from reconnecting to the greater whole of the earth. Identical developments are documented as an output of Somatic practice (in Chapter 4.)

Deep Ecology theory addresses Anthropocentrism (human centeredness) as a root cause of the environmental crisis. It wasn't until the mid nineteen seventies, with the emergence of Ecofeminism, that Androcentrism (male-centeredness) and the connection between man/nature domination and male/female domination were made.

In her 1975 book New Woman/ New Earth, which is widely acknowledged as being the first Ecofeminist book, Rosemary Ruether wrote: "Women must see that there can be no liberation for them and no solution to the ecological crisis within a society whose fundamental model of relationships continues to be one of domination. They must unite the demands of the women's movement with those of the ecological movement to envision a radical reshaping of the basic socioeconomic relations and underlying values of this society."[19] Following Ruether's statements, Ecofeminism has taken off in many directions, from the mystical to the political, with certain central strands with which most writers agree.

Ecofeminism's central premise is that the ideology which sanctions the oppression of nature, also sanctions the repression of women. Ecofeminists argue that the oppression of women and nature stem from the same root – an ideology that sanctions domination of man over nature. Many Ecofeminists, including Australian Ariel Kay Salleh, have critiqued Deep Ecology for its blindness to the connection between patriarchy and the domination of

---

[19] Ruether, Rosemary New Woman/ New Earth: Sexist Ideologies and Human Liberation (New York: The Seabury Press, 1975) 204

15

nature.[20] Philospher Ynestra King confirms that Ecofeminists "believe that the subordination of women in society is the root of human oppression, closely related to the association of women with nature." [21] The common ground with Deep Ecology is the seeking to redress a fundamental ideology of domination. Ecofeminism also offers a sound critique, in which the natural body as a problematic in the environmental crisis is clear.

According to prominent environmental philosopher Marti Kheel:

> There is a significant distinction between Ecofeminism and Deep Ecology, however, in their understanding of the root cause of our environmental malaise. For Deep Ecologists, it is the anthropocentric worldview that is foremost to blame....Ecofeminists, on the other hand, argue that it is the Androcentric worldview that deserves primary blame. For Ecofeminists it is not just "humans" but men and the masculinist worldview that must be dismantled from their privileged place.[22]

Deep Ecologists ostensibly refer to a gender neutral self, which Ecofeminists refute is inappropriate in a gender constructed world. Ecofeminists argue that it is not simply a dominant worldview that must be deconstructed, but the male-dominant worldview and world that must be taken on. As Zimmerman notes, "only when Deep Ecologists learn to

---

[20] Zimmerman, Michael, Contesting Earth's Future: Radical Ecology and Postmodernity. Pg. 9
[21] King, Ynestra "Healing the Wounds: Feminism, Ecology, and the Nature/Culture Dualism" in Reweaving the World: The Emergence of Ecofeminism Irene Diamond and Gloria Feman Orenstein Eds. (San Fransisco: Sierra Club Books, 1990) 109
[22] Kheel, Marti "Ecofeminism and Deep Ecology: Reflections on Identity and Difference in Reweaving the World: The Emergence of Ecofeminism Irene Diamond and Gloria Feman Orenstein Eds. (San Fransisco: Sierra Club Books, 1990) 129

appreciate the effects of patriarchal culture on their own awareness, only when they discover the extent to which their conceptions of self, body, nature and others are shaped by patriarchal categories, will their ecology become truly deep."[23] This is why Ecofeminists speak of Androcentrism in place of the anthropocentrism of Deep Ecologists. For Ecofeminists, it is meaningless for Deep Ecologists to try to address the dominant paradigm if they don't address the dominant relationship that built and supports it: patriarchy.

What Deep Ecology and Ecofeminsm share is a belief that the root of the problem is in the devaluation of nature, and the establishment of a human/nature dialectic. The prominent environmental author Charlene Spretnak writes, "the central insight of Ecofeminism is that a historical, symbolic, and political relationship exists between the denigration of nature and the female in Western cultures."[24] Spretnak does a superior job detailing the historically and philosophically connected oppression of nature and the female. According to Spretnak:

> From the Bronze age onward the denigration of nature and the female in European societies fluctuated but never disappeared. The Pythagoreans codified their influential table of opposites in which the female is linked with the negative attributes of formlessness, the indeterminate, the unlimited – that is dumb matter, as opposed to the (male) principles of fixed form and distinct boundaries. Aristotle considered females to be passive deformities. The

[23] Zimmerman, 142
[24] Spretnak, Charlene "Critical and Constructive Contributions of Ecofeminism" in Peter Tucker and Evelyn Grem Eds. Worldviews and Ecology (Philadelphia: Bucknell Press, 1993) 182

intellectual prowess of the male, he felt, could reveal and categorize all form and functions of organisms in nature. Later, the medieval cosmology ranked men above women, animals, and the rest of nature, all of which were considered to be entangled with matter in ways that the male spirit and intellect were not. The advent of modernity created by the succession of Renaissance humanism, the Scientific Revolution, and the Enlightenment shattered the holism (but not the hierarchical assumptions) of the medieval synthesis by framing the story of the human apart from the larger unfolding story of the earth community.[25]

While modern society seems to live in the world of Francis Bacon who said "the world is made for man, not man for the world," Deep Ecologists and Ecofeminists are seeking to re-pattern the entire relationship toward one of biocentric egalitarianism.[26] Of course this stands in the way of industrialists everywhere. But as the famous environmentalist David Brower,[27] put it: "I do not blindly oppose progress. I oppose blind progress".[28]

The past thirty years of research and writing in Deep Ecology have exposed the dominant relationship of man over nature as central to the environmental crisis. Ecofeminist authors have further developed the notion that the ideology that divides humanity and the natural body sanctions the

---

[25] Ibid, 183
[26] In The Ends of the Earth: Perspectives on Modern Environmental History Donald Worster Ed. (NY: Cambridge University Press, 1988) 20.
[27] Founder of the Sierra Club, and possibly best known because of John McPhee's biography of him, Encounters with the Archdruid
[28] Brower, David with Steve Chapple Let the Mountains Talk, Let the Rivers Run: A Call to Those Who Would Save the Earth. 78

domination of man over nature. Somatic Ecology argues that the dominant relationship of the Human to the natural world – as described by Deep Ecologists and Ecofeminists - is authorized by the separation of the Human from the Human body.

In his The American Replacement of Nature, William Irwin Thompson wrote:

> We can despair at this trashing of the world, and in our bitterness curse the age, for writers are good with words and will always find it easier to criticize and curse; but if we lift our imagination into the whirlwind that takes our home away, we might just see for a moment with the eye of the hurricane to experience a planetary atmosphere at work on the little things we ignored, and the big things we never knew.[29]

Deep Ecologists and Ecofeminists ask people to pay attention to the little things, and to strive to understand the big things. Somatic Ecology theorizes that the point of control for resolving the environmental crisis may be realigning our relationship our own natural selves. In order to accomplish this change, we must re-value Somatic Knowledge. In looking at the issue of toxic waste, or endangered species, it is easy to dismiss such an odd, and in a way, remote, solution. As Einstein noted, however, we must make things as simple as possible, and no simpler. The environmental crisis and its roots are truly complex.

---

[29] Thompson, William Irwin The American Replacement of Nature: The Everyday Acts and Outrageous Evolution of Economic Life (New York: Doubleday, 1991) 78

If our problem really is the dominant relationship of man over nature, and its roots really lie in the separation of humanity from nature – how might we find resolution? The following chapters will look at the Judeo-Christian tradition and the development of the Scientific Process to document the development of the Humanity/Nature dialectic. As we see that the Human/Nature dialectic has developed simultaneous to and within the separation of the Human from the Human body, a potential solution emerges for the re-placement of the Human within Nature.

Chapter Two: Understanding the Modern Body

"For I know that nothing good dwells within me that is in my
flesh.... Who will deliver me from this body of Death?"

- Romans 7:18-24

The Human/Nature dialectic has ancient roots, connected to the
evolving Human self-conceptualization. This chapter will focus on Judaism
and Christianity and how the body is regarded within those traditions. These
two religions have carried dominant influence in Western civilization; our
modern attitudes have been forged out of their understanding.

While both religions carry multi-faceted relationships to the body, this
chapter will focus on sex as justifying representative and compelling
conclusions regarding the body. The writings of the Jewish Kabbalah[30] will be
considered, alongside Christian Gnostic[31] texts and the writings of the Apostle
Paul.

It would be appropriate to offer the possibility that Christian attitudes
were more influential moving into the age of science (as will be examined in
the next chapter.) Understanding how prior centuries regarded the body
establishes the roots out of which grew the tree of modern science. As will be
seen through this analysis of the Jewish and Christian relationship to sex,

---

[30] Ancient Jewish spiritual scholarship.
[31] Early Christian mystics/spiritual scholars.

Humanity's relationship to the body has been troubled throughout Western history. It is this understanding which allows for accurate interpretation of the statement made in Job 19:26, "Yet in my flesh shall I see God."

The central components to the Jewish relationship to the body will be identified within an analysis of how Kabbalists constructed sexual reproduction, and further, the body as a spiritual metaphor. Reproduction was viewed largely as a positive act according to Modern Scholar Charles Mopsik. He notes that, "quite significantly, procreation has been compared to the Temple, the principal function of which was to bring the divine presence and His blessings into the world, like a captor of the divine forces vested in the cosmos: Rabbi Abin says, 'The Holy One, blessed be He, has greater affection for fruitfulness and increase than for the Temple.'"[32] The Temple, the symbol of the holiest place in the religion, is compared favorably to procreation. As will be explicated, in Judaism, "fruitfulness" is considered the most direct path toward to G'd.

The body in reproduction is constructed as a mirror of divine processes on earth. Again to quote Mopsik, "the human body as signifier is understood as the structural model of the divine cosmos."[33] The act of coupling is seen as a replication of the divine act of creation, and as such is connected to the spiritual world. The Kabbalist Rabbi Moses Cordovero quotes the Midrash (commentary on the Torah) in saying that "it is not without reason that the first instruction in the Torah is, 'Be Fruitful and Multiply.'"[34]

---

[32] Mopsik, Charles "The Body of Engenderment in the Hebrew Bible, the Rabbinic Tradition and the Kabbalah" in Fragments for a History of the Human Body Michael Feher ed. (New York: Zone, 1989) 57
[33] Ibid, 58
[34] Tefillah le-Moshe, p. 213 in Ibid, 60

The act of sex is seen as holy, and the body, as an instrument of G'd. Mopsik argues that the Jewish interpretation of sex as divine metaphor extends from Judaism's relationship of the human being to the divinity:

Kabbalistic sources lead us to the idea that through human engenderment the divinity accomplishes one more step in the process of its manifestation.... In order to move toward His fulfillment, in order to be personified, He must pass into time's texture woven by the thread of engenderments. Each new conception, each new birth is inscribed as an indispensable stage on the path that leads to divine manifestation, both eschatological and messianic. The body of engenderment is, therefore, a body of passage. Like the eye of a needle it allows the thread of theophanic becoming to move through time and weave its fabric. A thirteenth century Kabbalist, Rabbi Joseph de Hamadan illustrates this idea perfectly when he writes: 'He who has children extends as it were the existence of the chain of likeness which is the Chariot [the Divine]. Indeed, the latter is called the "chain of likeness"....He who is without children lessens the chain of likeness. Thus, every man who has children fulfills the Chariot on high.[35]

In Judaism, it is partly through the act of passing down the religion that the devotee connects to the divinity, and as such, it is through an act of

---

[35] Ibid, 61-2

embodiment that a person moves closer to G'd. Becoming a body – becoming embodied, and the act of embodiment - is conceived as a necessary step toward manifesting the deity on earth.[36] The person who engages in procreation is not simply helping in the act of creation but is fulfilling a *mitzvah* – a good deed - of the highest order, fulfilling the likeness of the Divine in human form. To summarize, sex is characterized as "eminently positive and constructive."[37]

A connected Kabbalistic notion is 'Sefirot' (plural of sefirah) According to Mopsik, "this word refers to the ten emanations issuing from En Sof, the ineffable Infinite, which form a spiritual structure in the shape of the human body."[38] This more arcane Kabbalistic structure, while not relating directly to sex, displays the extension of Kabblistic theory into epistemology. Sefirot places in the body the qualities (knowledge, wisdom, generosity, love, strength) that we find in modern times identified as qualities of mind. Images of Sefirot show strong parallels to images of spiritual energy channels within diverse eastern religions, including several connected to strong Somatic practices (including Yoga.) The visual duality of the spiritual within the human body represents the positive theoretical association by which the human body is a positive earthly carrier of the spiritual.

Kabbalistic writings on sex expose a worldview that values the body. Engenderment – sex - is seen as leading toward and being an act of Godliness. Embodiment is characterized as necessary and positive, in part for its

---

[36] This manifesting of G'd on earth through embodiment is carried to its highest level, of course, in Christianity, in which G'd himself took the form of a human. According to some Christian thinkers, this implies a positive relationship to the body – G'd was willing to be embodied.
[37] Ibid, 63
[38] Ibid, 58

connection to the continuation of the religion. Looking at Christian Gnostic writings, the sense is quite different.

Many facets of Christianity grew directly from Judaism – most simply the New Testament from the Old Testament. But regarding the relationship to the body, it appears that societal influences had a stronger impact than religious precedent. The Christian mindset appears to grow from Greek conceptions of the body. The relationship between Greek and Christian understandings is exhibited broadly, including in the following from Clement of Alexandria: "the human ideal of continence, I mean that which is set forth by the Greek philosophers, teaches one to resist passion, so as not to be subservient to it, and to train the instincts to pursue rational goals. [But as Christians] our ideal is not to experience desire at all."[39] While many are familiar with the celebrated sensuality of Greek culture, there was simultaneously an isolation of control over the body from within the human conception. As Gnostic writings make clear, early Christians take the isolation of the spirit, exhibited in the Greek ethos, to extremes. According to Greek scholar Jean-Pierre Vernant:

> The body is the agent and instrument of actions, powers and forces which can only deploy themselves at the price of a loss of energy, a failure, a powerlessness caused by congenital weakness....But it is always Death, in person or by delegation who sits within the intimacy of the human body like a witness to its fragility. Tied to all the nocturnal powers of confusion, to a

---

[39] Clement, Stromateis, in Henry Chadwick, trans. "Alexandrian Christianity", p.66 in Brown, Peter The Body and Society: Men, Women and Sexual Renunciation in Early Christianity (New York: Columbia University Press, 1988) 31

return to the indistinct and unformed, Death, associated with the tribe of his kin – Sleep, Fatigue, Hunger, Old Age – denounces the failure, the incompleteness of a body of which neither its visible aspects... nor its inner forces of desire feeling thoughts and plans are ever perfectly pure... Thus for the Greeks of the archaic period, man's misfortune is not that a divine and immortal soul finds itself imprisoned in the envelope of a material and perishable body, but that his body is not fully one. It does not possess, completely and definitively, that set of powers, qualities and active virtues which bring to an individual being's existence a constant, radiant, enduring life in a pure, totally alive state, a life that is imperishable because it is free from any seed of corruption and divorced from what could, from within or without, darken, wither and annihilate it.[40]

The Greek difficulty with the body rested on its inherent divisions and weakness, divisions and weakness which separate the human body/humans from godliness. Clearly this is in opposition to the Jewish characterization of the human/deity relationship. As in the Greek understanding, Christian writings make a tie between the weakness of the body, and the body as impediment to a higher spiritual calling.

This is visible in the Apostle Paul's famous Corinthian Letter. The letter responds to the community in Corinth, which was agitating to create a Utopic society in preparation for the coming of Christ. As Peter Brown establishes in

---

[40] Vernant, Jean-Pierre "Dim Body, Dazzling Body" in Fragments for a History of the Human Body Michael Feher ed. (New York: Zone, 1989) 25

is brilliant text <u>The Body and Society: Men, Women and Sexual Renunciation in Early Christianity</u>, the Corinthians proposed a radical ideal.

> [The Corinthians] would undo the elementary building blocks of conventional society. They would renounce marriage. Some would separate from pagan spouses; others would commit themselves to perpetual abstinence from sexual relations. The growing children for whose marriages they were responsible would remain virgins. As consequential as the Essenes, they would also free their slaves. Somewhat like the little groups described by Philo outside Alexandria, men and women together would await the coming of Jesus 'holy in body and spirit'.[41]

At a time when Christianity was just growing, the Corinthians radical notions threatened the inclusion of more mainstream elements, and so Paul wrote to put down this rebellion. That a critical concept within the religious fringe was abstinence is telling. That Paul himself was celibate points directly to early Christianity's troubled relationship to the body.

In ministering the Corinthians toward sex, his words expose a very negative conception of the act. In Corinthians 7:36-38 Paul wrote:

> If any one thinks that he is not behaving properly toward his betrothed [in some versions 'virgin'], if his passions are strong, and it has to be, let him do as he wishes; let them marry – it is no sin. But whoever is firmly established in his heart, being under

[41] Brown, Peter <u>The Body and Society: Men, Women and Sexual Renunciation in Early Christianity</u> (New York: Columbia University Press, 1988) 52

no necessity but having his desire under control, and has determined this in his heart, to keep her as his betrothed, he will do well. So that he who marries his betrothed does well; and he who refrains from marriage will do better.[42]

Paul declares that marriage is a negative, undertaken only to ward off the sin of sex before marriage. Second, marriage, and sex, are negatives that are better to be refrained from altogether. Paul states this even more clearly in an earlier passage, Chapter 7 verses 32-34.

I want you to be free from anxieties. The unmarried man is anxious about the affairs of the Lord, how to please the Lord; but the married man is anxious about worldly affairs, how to please his wife, and his interest are divided. And the unmarried woman or girl is anxious about the affairs of the Lord, how to be holy in body and spirit; but the married woman is anxious about worldly affairs, how to please her husband.[43]

Marriage calls into question the ability to focus on the Lord. Married people lack the quality of what Brown analyzes as "the undivided heart", and are therefore lesser Christians than those who are married solely to G'd.

---

[42] Paul in The Writings of St. Paul ed. Wayne A. Meeks (New York: W.W. Norton and Company, 1972) 34
[43] Ibid, 33

Paul does not himself have a problem with celibacy. He believes that the body and bodily attachments impede connection to G'd.[44] This is, again, in direct contrast to the Jewish tradition. As Jewish scholar Blu Greenberg writes, "In Jewish tradition, one who doesn't marry is considered an incomplete person. Marriage, not celibacy is the higher form of existence. It is no coincidence that the word for marriage, *Kiddushim*, is derived from the root word *Kaddosh*, holiness."[45]

The complex divide between the physical and spiritual human is further exposed in Corinthians 6:20: "You are not your own; you were bought with a price. So glorify G'd in your body."[46] This exhortation seems at first strikingly out of place. To glorify G'd in the body according to the succeeding passages is to deny the body. There is therefore a dualism in Paul's words. "You are not your own" implies a slave/master relationship, replicated in the body/spirit relationship which Paul calls for in relationship to the body. This relationship seems to have parallels with the Greek conception.

The line, "You were bought with a price" furthers the imagery of the Christian body as a slave. Purchase of the Christian human, and resultant proximity to G'd, was made with the death of the Messiah, with the death of Christ. The social order at the time made this language plain, though today it may seem convoluted. In application, how were Christians, as slaves, to glorify G'd in their body? The preceding passages clarify the body as a part of humanity that must be denied to approach holiness. In light of all of the prior evidence it would seem that 'glorifying G'd in their bodies' was a statement intended to discourage one from indulging in any acts which would lead to a

---

[44] Paul states in 7:7, "I wish that all were as myself am." (Paul was married but became celibate and left his wife after receiving the call to higher purpose.)
[45] Greenberg, Blu How To Run a Jewish Household (Northvale: Jason Aronson, 1989) 216
[46] Paul, 32

divided heart. In order to prove oneself to the Lord, what is intended is the isolation of the human body, from the spiritual body. The Corinthian letter furthers the notion of the human body, and its human sensations, as requiring isolation from the best of the human – the spiritual human.

There are other images in Christian Gnostic writings, similar to the human/body slave imagery, which represent a troubled relationship to the physical human. In the Apocryphon of John (the Secret Book of John) from the Nag Hammadi manuscripts, comes this image of the body as prison:[47]

And I entered the midst of their prison,

Which is the prison of the body.

And I said, 'O listener, arise from heavy sleep.'

And that person wept and shed tears, heavy tears;

And wiped them away and said, 'Who is calling my name?'

'And from where has my hope come, as I dwell in the bonds of the prison?'[48]

The Book of Thomas the Contender, another Nag Hammadi manuscript, also contains a prison reference, "Woe to you who put your hope in the flesh and the prison that will perish." The contrast to the Jewish tradition is at times direct. In the Jewish tradition sexual intercourse is next to godliness. In some Christian Gnostic teachings sex is associated with the

---

[47] The Nag Hammadi manuscripts were found in Upper Egypt in 1945 and contain fourth century A.D. copies of many writings from various Gnostics. Greek originals from some of the works were probably composed as early as the second century A.D. and a few even earlier.
[48] Apocryphon of John, NHC 2.31..3-9; Layton, The Gnostic Scriptures, p.326 in Williams, Michael A. "Divine Image – Prison of Flesh: Perceptions of the Body in Ancient Gnosticism" in Fragments for a History of the Human Body Michael Feher ed. (New York: Zone, 1989) 136

31

animal world, and the fall from a true path. Gnostic scholar Michael Williams writes, "in the Book of Thomas the Contender, the human body is said to be something that is 'beastly', which will perish like the bodies of beasts, and which can never beget anything different from what beasts beget, since it itself was produced through sexual intercourse, just like the bodies of beasts are produced."[49] In contrast, as Rabbi Abraham Witty writes, "In Judaism, human sexuality is raised to a level of sanctity above the mere satisfaction of the biological and psychological urges associated with it."[50] It is actually a mitzvah (a good deed) to have relations as a married couple, and a *double mitzvah* if you have relations on Shabbat (between sundown Friday and sundown Saturday.) While being in the continuum of existence in the Kaabalistic tradition is a positive, within the Christian ideology it represents the distance between the holy and the living. This simple conceptualization of life, and its propagation, in opposition to spiritual existence is striking in its negativity.

The story of Christ's Immaculate Conception is important here as he was born, according to the story, outside of the beastly means. And just as Christians strive to be Christ-like, they are bound trying to transcend their physical form, which imprisons them and ties them. Michael Williams writes, "according to Hippolytus of Rome, some second century Gnostics, whom we call Naasenes, taught that intercourse was something appropriate for pigs and dogs rather than humans."[51] Mopsik adds to this:

---

[49] Williams, Michael A. Divine Image – Prison of Flesh: Perceptions of the Body in Ancient Gnosticism in Fragments for a History of the Human Body Michael Feher ed. (New York: Zone, 1989) 138
[50] Witty, Rabbi Abraham B. and Rachel J. Witty Exploring Jewish Tradition (New York: Doubleday, 2000) 412
[51] Ibid, 138

If he recognizes that he owes his birth to the desire that his progenitors mutually gave themselves to, a mutual desire awakened by the law of fidelity, he is inscribed effortlessly in the law of engenderments. He becomes a unique moment in the process of creation which he in turn will extend still further. The Pauline split turns precisely on this point: spiritual Israel is instituted beginning with the Christology that broke the chains of births: Christ's father is not His mother's spouse......The inevitable consequence is that the individual body is no longer the mirror in which the bodies of preceding generations converge and are reflected....and henceforth in Paul's doctrine, this breach opens a gulf between what is known as the carnal relation and the spirit that has been delivered of the flesh, that is, from what is reproduced along family lines.[52]

Paul's Corinthian letter expressed a sense that the body was something that should be transcended. Many Gnostics writings align with this philosophy, detailing a clear aversion to the sensate body. Where Paul was celibate and saw his body as a slave, the Gnostics abhorred the need to be beastly. The contrasts to the Jewish system could not be more stark. One of the most famous Rabbis, the Zohar wrote, "Where there is no union of male and female, men are not worthy of beholding the Shekhinah, the Divine Presence."[53]

There are other aspects within the Jewish tradition that show a more ambiguous relationship, in particular a relationship which sets the roots for

---

[52] Mopsik, 55
[53] Greenberg, 120

what the Ecofeminists identify as Androcentrism – control of the female by the male. These roots are exhibited in the laws of Taharat HaMishpachah, and the practice of the Sheitel. The Sheitel is the name for the wig which all married Jewish women must wear, if they do not wear a scarf or other form of head covering, after they have shaved their head. The head is shaved because nobody but the man whom the woman marries is supposed to see the woman's hair after marriage. As George Robinson writes in Essential Judaism, "throughout the debate on men's head covering, all the sages were in agreement on one thing: married women must cover their hair. Even in Biblical times, it was considered a brazen violation of the rules of modesty for a married woman to allow anyone but her husband to see her hair."[54] By requiring that women shave their heads, this Jewish tradition turns the body into a site of repression.

Cynthia Baker, in her article on the domestication of Jewish bodies, appropriately highlights the Talmudic Rabbinic maxim regarding the relative visibility of women when not at home – be they uncovered or covered. "A woman who has a husband: whether she adorns herself or not, people do not stare at her. And if she does not adorn herself, she will be despised.... A woman who has no husband: whether she adorns herself or not, everybody stares at her."[55] The laws requiring that women cover their heads, or shave their heads, places women's bodies in a controlled environment. While the acts of procreation are represented positively in Judaism, the control of the body still exists – exhibited within a patriarchal context.

---

[54] Robinson, George Essential Judaism: A Complete Guide to Beliefs, Customs and Rituals (New York: Pocket Books, 2000) 28-9
[55] T. Qiddushin 1:11 in Baker, Cynthia M. "Ordering the House: On the Domestication of Jewish Bodies" in Parchments of Gender: Deciphering the Bodies of Antiquity ed. by Maria Wyke (Oxford: Clarendon Press, 1998) 239

The laws of Taharat HaMishpachah are also known as the Laws of Family Purity. They cover the rules surrounding sexual relations around the time of a woman's monthly menstrual flow. According to the laws, which are laid down not once, but three times (Leviticus 15:19, 18:19, and 20:18) a woman is in Niddah, or ritual impurity, for the time of her period and for seven days after the flow has stopped. She continues to be impure until she has taken of the Mikveh, a purifying bath.[56] To turn a natural bodily function into a state of religious impurity is to place the body into a negative cultural space. To equate 'family purity' with labeling the wife impure for a significant portion of every month is both placing a negative connotation on a natural bodily function and blowing up that negative association into a belief that holds huge implications for others (family purity.) The relationship to the body in Judaism is overwhelmingly one of respect, but in relationship to the female body are strong ties to Androcentrism.

The Jewish body in reproduction is a binding metaphor for creation that ties believers to divinity. Further than that, knowing the body is more than just a metaphor for seeking knowledge of the divine, as can be seen in writing on the Sefirot. As Rabbi Moses Hayim Louzatto states, "in other words, the relationship among Sefirot, the law of their functioning, is the same as the law of the functioning of the body in all its parts. From this comes the understanding of the verse 'Yet in my flesh shall I see G'd', in order to see and comprehend all the doings of man and the whole of his movements, all of which have their roots deep in the Sefirot."[57] Respect for the body replicates the process of studying theology. Body wisdom takes on a whole new

---

[56] Witty and Witty, 412-413
[57] Louzatto, Moses "Addir ba-Marom", Jerusalem, 1968, 2a in Mopsik, 65

neaning from this perspective, applying to a spiritual the act of procreation, and every day re-creation of the body in life/death.

On the opposite extreme there is Paul and the Christian tradition. In Romans 7:18 he writes:

> For I know that nothing good dwells within me, that is, in my flesh. I can will what is right, but I cannot do it. For I do not do the good I want, but the evil I do not want is what I do. Now if I do what I do not want, it is no longer I that do it, but sin which dwells within me. So I find it to be a law that when I want to do right, evil lies close at hand. For I delight in the law of G'd, in my inmost self, but I see in my members another law at war with the law of my mind and making me captive to the law of sin which dwells in my members. Wretched man that I am! Who will deliver me from this body of death? Thanks be to G'd through Jesus Christ our Lord! So then, I of myself serve the law of G'd with my mind, but with my flesh I serve the law of sin.[58]

This attitude - fear, and hatred of the body - is characteristic of a Christian tradition that sets the spiritual body apart from the living body. By applying passion, desire, and sin, to the living body the Christian tradition encourages a schism from the body as a seat of knowledge. The direct result of this morality is that the connection between body and mind became more distant, and a wider cultural project to validate the distance was enabled. The

---

[58] Paul, 79-80

Jewish tradition, in it similarities to other traditions which set the body as model for a whole presence on the earth, remains a potential model for a healthier link to the body in the future.

As we shall see in the following chapter, the Christian body/mind division and fear of the body was enshrined in epistemological efforts amongst emerging modern Western scientists and philosophers, who fought to understand and control the world with a post-religious mindset. The conceptual division of the human from the body in spiritual theory sanctioned the full divide of human from human/nature (the body) in the development of modern science. The development of modern epistemology built on the existing body/mind problematic, validating a dominant relationship between humanity and the greater ecology.

Chapter 3: The Rise of Disembodied Reason - Galileo, Descartes, and Locke

This chapter will focus on three important thinkers in science and philosophy whose views have been seminal in leading to our perception of the body in the twentieth and twenty-first centuries. Galileo is one of the founders of modern science and as such helped construct a relationship to Knowledge that is mathematical, reductionistic and technological. Descartes is considered by many the founder of the modern conception of the self; his epistemology cemented a body/mind schism. Locke is the leader of the Empiricists, a group of philosophers who used reason to re-introduce the body as a tool for Knowledge. Together, with others, these three figures built the modern human construction, enshrining a body/mind split that simultaneously cemented a distance between humanity and nature. What was once a moral stance – as we saw in chapter 2 – became a logical platform in the sixteenth to the eighteenth centuries.

Galileo Galilei (1564-1642), was the first son of an accomplished musician, Vincenzio Galilei.[59] Galileo dropped out of college in 1585, having

---

[59] Galileo, who would spend his entire lifetime fighting for objectivity, was born to a family which supported questioning and intellectual rigor over faith in tradition. Galileo's father in particular clearly influenced his intellectual bent. Consider the following writing, from a commentary in his father's Dialogue of Ancient and Modern Music which was published at the time that Galileo was in University. "It appears to me that they, who in proof of any assertion rely simply on the weight of authority, without adducing any argument in support of it, act very absurdly. I on the contrary, wish to be allowed freely to question and freely to answer you without any sort of adulation as well becomes those who are in search of truth." (in Fermi, Laura and Bernardini, Gilberto Galileo and the Scientific Revolution (New York: Basic Books, 1961) 8

abandoned a medical course and taken up math and philosophy, but completing degrees in neither. In the small score of years following his exit from university he began his experiments, writing a treatise on motion from which he later drew his greatest contributions to physics. Following an address to the Florentine Academy in 1588 on the location, size and arrangement of hell as described in Dante's inferno (which was particularly exacting in its mathematics), the head of that society helped Galileo get a position as a math professor at the University of Pisa, and subsequently at Padua.

Galileo's important work and his history will be summarized, so that an understanding of his methodology of Knowledge development through reductionism, and the use of mathematics and machines, can be made clear. The success of his focus resulted in taking Knowledge generation away from the body, and placing it instead into the mind and its creations. As Galileo scholar Peter Machamer writes "this is not to say that Galileo had no predecessors. Of course he had many. But no one, until Galileo, made the mechanical way, the way of doing science, the way of knowledge."[60] While Galileo is best remembered for leading the charge for Copernicanism (the ideas for which he was punished by the Church), it is his method of science, a method that used machines and mathematics, that is most pertinent to modern epistemology. Galileo's incredible success using the telescope and mathematics created a separation from the body as a means to achieve knowledge.

It is impossible to fully understand the changes that occurred in the sixteenth and seventeenth century without noting the system against which

---

[60] Machamer, Peter "Galileo's Machines, His Mathematics, and His Experiments" in The Cambridge Companion To Galileo ed. Peter Machamer (Cambridge: Cambridge University Press, 1998) 54

these scientists and philosophers were working. The Aristotelian system of beliefs was the backbone of science and philosophy leading up to the historical point in question. According to historian of science Stillman Drake "Aristotle was, according to Dante 'the Master of those who know' so regarded by learned men from the time of Aquinas to that of Galileo. If one wished to know, the way to go about it was to read the texts of Aristotle with care, to study commentaries on Aristotle in order to grasp his meaning in difficult passages, and to explore questions that had been raised and debated arising from Aristotle's books... The Principles of physical science were determined in Aristotle's Metaphysics."[61] The main points were as follows, the earth was the center of the universe; the sun and stars were perfect and unchangeable and moved in perfect circular motion; each area of inquiry could be dealt with in a unique fashion, with a unique method; logical rules could be established by which causes could be determined for the effects we see in nature; four elements and four qualities in paired opposites comprised all matter. The arguments of Galileo and to some extent Descartes were largely asserting that Aristotelian natural philosophy was simply not true; that it was not good science. This resulted in a major change in philosophical thought.

An early example of Galileo's experimental work is his law of falling bodies, which he discovered in 1603. Galileo set up an experiment in which he let a ball roll down a sloping plane, noting its position after a series of equal times, judged by musical beats. By extrapolating mathematically, he was able to develop the law. His success meant that his method would be copied. And his method was based on a mathematical, mechanistic, approach

---

[61] Drake, Stillman <u>Galileo</u> (New York: Hill and Wang, 1980) Pg. 8

41

o the world. In his Il Saggiatore in 1623 Galileo wrote: "Philosophy is written in this grand book – I mean the Universe – which stands continuously open to our gaze, but it cannot be understood until one first learns to comprehend the language in which it is written. It is written in the language of mathematics, and its characters are triangles, circles, and other geometrical figures, without which it is humanly impossible to understand a single word of it; without these one is wandering about in a dark labyrinth."[62] Galileo made experiment (largely with the aid of machines) and mathematics the prime means for the discovery of knowledge, as can be seen from the law of falling bodies. Galileo made observations both objective and quantitative, and his future work, and fame, centered on extending the power of the human senses beyond their normal physiological limits. The significance of his method was that it further removed the body from the direct process of innovation, inserting instead the mind as the most trenchant facilitator in discovery.

In 1609 Galileo was informed about a device that made objects appear closer than they were and realizing its potential importance to Venice, a port city, he decided to work on developing such an instrument. Shortly after the telescope's successful debut, Galileo set his sights on the heavens. He discovered first that the moon was not perfectly round – rebuking another belief held by Aristotle. Galileo then found that there were an uncountable number of stars, and in early January of 1610 he found four planets around Jupiter. All of this he detailed in the book A Starry Messenger (1610.) His

---

[62] Galilei, Galileo Il Saggiatore ((Milano:Feltrinelli, 1915, 1623) 38 in Machamer, Peter "Galileo's Machines, His Mathematics, and His Experiments" in The Cambridge Companion To Galileo ed. Peter Machamer (Cambridge: Cambridge University Press, 1998) 62

fame from the book led to his appointment as court mathematician and philosopher to Cosimo de Medici, ruler of Florence.

Galileo's discoveries were the subject of discussion throughout Europe within weeks of the time he made them. As soon as Johannes Kepler read The Starry Messenger he had it translated into German, and five years later it appeared in China in a Chinese translation (a prompt turnaround in those days.) Some didn't believe his findings at first, given that one had to have the proper instruments to verify the discoveries and only Galileo was capable of constructing the necessary technology. However, Galileo slowly won over his doubters and Kepler, after being delivered a Galilean telescope and pointing it at the heavens wrote to Galileo "Galileae Vicisti" (You have won, Galileo.) [63]

Galileo was feted in Rome in 1611 and had an audience with the Pope, who was interested in his discoveries. A return trip in 1616 however did not go as well. In late February of that year, Galileo's friend Cardinal Bellarmine, on the Pope's instructions, instructed Galileo not to hold, teach or defend the opinions of Copernicus (that the earth moved around the sun), which naturally followed from his discoveries in The Starry Messenger, and subsequent work. As Galileo scholars Fermi and Bernardini recount it, "a few days later the second event took place: years after its publication, the book De Revolutionibus, which Canon Copernicus had dedicated to a Pope, which the Pope had accepted, and which the Church had found no fault in until Galileo had started to present it as a reality, was condemned and prohibited until it should be corrected."[64] The true problem, again according to Fermi and Bernardini, was that "At the hands of Galileo... God's creation was becoming

[63] Fermi, Laura and Bernardini, Gilberto Galileo and the Scientific Revolution (New York: Basic Books, 1961) 63
[64] Ibid, 83

n object of direct human observation, which could be interpreted without the help of Scripture or of religion."[65] Galileo was mute for a number of years. But in the spring of 1624 the new Pope, Urban VIII (who was in fact Galileo's old friend Cardinal Barberini), agreed that he could write about Copernicus' system as long as he represented it not as reality but as scientific hypothesis. Galileo was sixty now and in poor health. It took him four years to write his <u>Dialogue Concerning the Two Chief World Systems.</u>

Following the publication of the book in 1632, the Pope ordered Galileo be brought to Rome and tried by the Inquisition. The Inquisition stated that "the author claims to have presented a mathematical hypothesis, but he gives it a physical reality, something a mathematician would never do."[66] Found 'suspect of heresy' he was ordered to recant, which he did, and was sentenced to imprisonment, which was commuted to house arrest for the remainder of his life (till 1642.) In a letter to Nicole Fabri de Peiresc, a French amateur scientist, Galileo wrote:

> I have said that I hope for no relief, and this is because I have
> committed no crime. If I have erred, I might hope to obtain
> grace and pardon; for transgressions by subjects are the means
> by which the prince finds occasion for the exercise of mercy and
> indulgence. Hence when a man is wrongly condemned to
> punishment, it becomes necessary for his judges to use greater
> severity in order to cover up their own misapplication of the
> law. This afflicts me less than people may think possible, for I

---

[65] Ibid, 84
[66] In Rossi, Paolo <u>The Birth of Modern Science</u> Translated by Cynthis de Nardi Ipsen (Blackwell: Oxford University Press, 2000) 93

have two sources of perpetual comfort – first, that in my writings there cannot be found the faintest shadow of irreverence towards the Holy Church; and second, the testimony of my own conscience, which only God in Heaven thoroughly know. And He know knows that in this cause for which I suffer, though many might have spoken with more learning, none, not even the ancient Fathers, have spoken with more piety or with greater zeal for the Church than I.[67]

Galileo asked for, and was granted, special dispensation to attend church while under house. He did not find his science to threaten his religious philosophical understandings. Galileo's overthrow of Aristotelian metaphysics did, over time, result in the development of a modern metaphysics that challenges some religious canon. This modern metaphysics understands the human at a distance from the rest of the natural world.

Galileo's use of mechanical instruments and reliance on mathematics were both successful, and epoch-making. The most striking feature of Galileo's science, according to history of philosophy analyst Ernan McMullin was its emphasis on Mathematization.[68] Despite the fact that he ended his life under house arrest, according to Italian science historian Paolo Rossi, "Galileo had won, because in order to win over the final, unyielding objectors and silence the professors who rejected his discoveries on the grounds of logic and metaphysics, not even 'the sworn testimony of the stars themselves would have sufficed,' as he later wrote. The reality of the universe had been

---

[67] Drake, 92-3
[68] McMullin, Ernan "Galileo Galilei" in Routledge Encyclopedia of Philosophy Gen. Ed. Edward Craig (New York: Routledge, 1988) 837

xpanded through the use of a mechanical instrument..... for the men of his
ay Galileo's astronomical observations not only signified the end of one
world view but represented the birth of a new concept of experience and
ruth."[69]

The implications of Galileo's work from a sociological history of the
body, are that Galileo's experiments and writings helped eliminate the body
as a direct experiencer and producer of knowledge. The knowledgeable
human was no longer connected to the human body. Scientists following in
Galileo's footsteps continued to emphasize the mind, through mathematics,
and instruments, through technology, in the production of Knowledge. The
body as a direct, empathic source of Knowledge would recede in the re-
developing human conception. This was further enshrined in the work of the
philosopher/scientist Rene Descartes, who would carry Galileo's reason and
dispassion further – applying them skeptically, and with a resultant logical
severing of body and mind.

Rene Descartes (1596-1650) was born in Touraine, France and raised
by a lawyer/judge father and a nurse, his mother having died when he was
one. In 1616 he received his law degree from the University of Poitiers, but
shortly thereafter an inheritance allowed him to give up the law, and he set
out to travel the world. In 1618 he met a scientist named Isaac Beeckman,
who was to become a lifelong friend and mentor. Beeckman introduced
Descartes to an Anti-Aristotelian, mechanical, reductionistic approach to
nature.

---

[69] Rossi, 49

46

In 1619 Descartes had a series of three dreams, which he believed wer divinely sent, that he said inspired his search for a unified science an philosophy. He continued to travel for several years, undertaking variou experiments, including some on light and motion. It was during this time tha he wrote his first paper, the "Compendium Musicae", a music theory tract, fo Beeckman. In 1628 Descartes settled in Holland, where he would stay fo twenty years.

By 1633 he had finished <u>Le Monde</u>, his work of unified scienc combining the study of light, optics, meteorology, physics and biology Popular science author Daniel Boorstin recounts:

> Just as he was about to send off the manuscript to a friend he had a piece of shocking news. He had read Galileo and shared his Copernican Helio-Centric news. He learned that all copies of <u>Dialogue of the Two Chief Systems</u> had been burned and the author sentenced to an indefinite term of imprisonment. To Mersenne, Descartes wrote, 'I was so astounded that I quasi resolved to burn all my papers or at least not to show them to anyone. I cannot imagine that an Italian, and especially one well thought of by the Pope from what I have heard, could have been labeled a criminal for nothing other than wanting to establish the movement of the Earth. I know that this has been censured formally by a few cardinals, but I thought that since that time one was allowed to teach it publicly even in Rome. I confess that if this is false, then, all the principles of my philosophy are false also…. And because I would not want for

anything in the world to be the author of a work where there was the slightest word of which the church might disapprove, I would rather suppress it altogether than have it appear incomplete – "crippled", as it were.' [70]

Four years of his work were gone (Le Monde would be published posthumously), and it is interesting to note that in the future Descartes' work would be on topics (mathematics, physics, and optics and of course philosophy) that could not get him into trouble with the church. Descartes invented much of the basic vocabulary of algebra and geometry, including the form of the equation, the use of a and b for knowns, of x and y for unknowns, of numerals to express powers, and the form of the square root sign.

Though Descartes had earlier worked for a number of years on his Rules for the Direction of the Mind, it wasn't until his Discourse on Method (1637) and Meditations on First Philosophy (1641) that his full philosophical/scientific program was laid out. What Galileo did for science – forcing it to conform to mathematical principles – Descartes was able to do for epistemology. He created a system of logic, based on the foundations of mathematics, which serves as the basis for our modern understanding of the self, and our ways of knowing.

Descartes was looking for certainty, and as he says in the Discourse "only Mathematicians have been able to find any demonstrations, that is to say, certain and evident reasonings."[71] His claim was that he developed a method of demonstrating truths according to the exigencies of reason itself.

---

[70] Boorstin, Daniel J. The Seekers: The Story of Man's Continuing Quest to Understand His World (New York: Random House, 1998) 142
[71] Descartes, Rene Discourse on Method, (Indianapolis: Hacket Publishing Company, 1998, 1637) pg. 11

His question was, 'how can I know anything absolutely certainly?' His method made use of the method used to develop the only certain knowledge he knew: mathematics. He wished to make truth conform to a rational scheme. To examine philosophical issues mathematically, he created a mechanistic account of material reality. His mechanism and reductionism can be seen in the method outlined in the Discourse, "(I will) divide each of the difficulties I would examine into as many parts as possible and as was required in order to better resolve them."[72]

But simply dividing the problems into many parts was not sufficient to seek the truth – Descartes made doubt the beginning of his philosophy and certainty the first principle of his method. He wished to prove metaphysical arguments in the same fashion as Galileo proved astronomical ones. Not inconsequentially given the societal pressures, the idea he chose to prove in the Discourse was the existence of God. As noted philosophical historian Frederick Copleston writes, "What he was seeking was not to discover a multiplicity of isolated truths but to develop a system of true propositions, in which nothing would be presupposed which was not self-evident and indubitable."[73] Descartes' interest was in tracking the operations of the mind by which one is able, entirely without fear of illusion, of arriving at true knowledge of things. He begins by rejecting any opinion that might be called into doubt – and this quickly includes everything he once knew, because the senses cannot be trusted since "it is prudent never to trust completely those who have deceived us even once."[74]

---

[72] Ibid, 11
[73] Copleston, Frederick A History of Philosophy Vol. 4 (Westminster: The Newman Press, 1961) 66
[74] Descartes, Rene Meditations on First Philosophy (Indianapolis: Hackett Publishing Co., 1641; 1998) 60

Descartes tries to assess what he truly is, in his most basic form, and decides that the one thing he can know for certain is that he is a thinking thing.[75] But there must be more. "What else am I? I will set my imagination in motion. I am not that concatenation of members we call the human body. Neither am I some subtle air infused into these members, nor a wind, nor a fire, nor a vapor, nor a breath, nor anything I devise for myself...... But what then am I? A thing that thinks."[76] Descartes follows in the Christian tradition of rejecting the body, but for different reasons. Descartes reductionism is so extreme that he has reduced human existence to its very smallest front – to consciousness of one's self as a thinking thing.

Descartes also assigns emotion to the body, noting "bodily tendencies toward mirth, sadness, anger, and other such affects."[77] If reason lives in the brain, then emotion lives in the body. Descartes concludes that the body, while being a distinct idea, is not necessary.

For this reason, from the fact that I know that I exist, and that at the same time I judge that obviously nothing else belongs to my nature or essence except that I am a thinking thing, I rightly conclude that my essence consists entirely in my being a thinking thing. And although perhaps (or rather, as I shall soon say, assuredly) I have a body that is very closely joined to me, nevertheless, because on the one hand I have a clear and distinct idea of myself, insofar as I am merely a thinking thing and not an extended thing, and because on the other hand I have a distinct

---

[75] Ibid, 65
[76] Ibid, 65-66
[77] Ibid, 94

50

idea of a body, insofar as it is merely an extended thing and not a thinking thing, it is certain that I am really distinct from my body, and can exist without it.[78]

Just as Galileo divided problems into smaller units so as to be able to better examine them, so did Descartes.[79] It is interesting to consider, however that Galileo's mechanism is taken further with Descartes' ideas. While Galileo introduced machines as necessary adjuncts in the search for truth Descartes applied mechanism to the search for the self, and human truth.

And a clock made of wheels and counter-weights follows all the laws of nature no less closely when it has been badly constructed and does not tell time accurately than it does when it completely satisfies the wish of its maker. Likewise, I might regard a man's body as a kind of mechanism that is outfitted with and composed of bones, nerves, muscles, veins, blood, and skin in such a way that, even if no mind existed in it, the man's body would still exhibit all the same motions that proceed either from a command of the will, or consequently, from the mind.

Descartes seems to be talking about man as a robot, which is significant in that he is considered by many to be not only the father of modern

---

[78] Ibid, 96

[79] It should not be underestimated how powerful a notion the elimination of the body as a necessary adjunct to life must have been. Christians had been lamenting the body for over a thousand years. Now along came a philosopher who not only proved the existence of God (see Meditation Three – On God, That He Exists) but also got rid of that annoying stone around one's philosophical neck, the body. And Descartes' succeeded in doing so through the use of reasoning, which reduced problems to their smallest components, and through applied mathematical principles.

hilosophy, but also the creator of the modern idea of the self.[80] That this self s separate from the body, that it does not need the body, is of extraordinary mportance. Knowledge, from the mid-seventeenth century on, would be onstructed as something that is an aspect of the mind alone. Those particles f knowledge which cannot be confirmed by reason, objectively, that are ssociated with empathy, would be stashed away in the body and discounted rom the seventeenth century on.[81]

Despite Descartes' attempts to stay away from trouble, his philosophy vas seen as radical, and it found a lot of opposition. His entire opus was anned by the Universities of Utrecht and Leiden in the 1640's, and Cartesian hilosophy was condemned in 1656 in all the Netherlands by the Decree of he Synod of Dordrecht. In 1663 the Catholic Church listed Descartes' published works on the Index of Prohibited Books. However, by the late seventeenth century all the great universities of Europe had accepted Descartes' method and the bans against it had lapsed.[82]

Descartes' structure is based on disembodied reason. This provides for a mechanized understanding of the world and was applied successfully to all the sciences. The immense success of its application guaranteed its place in the pantheon. Nevertheless, the fact that it used extreme doubt toward the senses opened the door for a challenge by future philosophers who would wish to find a means to construct knowledge based on the world as experienced by the senses.

---

[80] Boorstin, 139

[81] Descartes at the end of his Meditations notes "hence I should no longer fear that those things that are daily shown me by the senses are false. On the contrary, the hyperbolic doubts of the last few days ought to be rejected as ludicrous." This admission, however does little in the way of revising the powerful logic which he has spent the previous days constructing, and is not an actual argument for embodied knowledge.

[82] Rossi, 99

John Locke (1632-1704) was born at Wrington in England, an educated at Oxford where he received his B.A. and M.A. Subsequently h became a lecturer in Greek and later Reader in Rhetoric and Censor of Mora Philosophy, still at Oxford. In 1666 he met Lord Ashley, later First Earl o Shaftesbury, a leading figure at the court of Charles II. A year later he joine the Earl's household, and for the next fourteen years shared in the fortune and misfortunes of Ashley, serving in a number of supportive bureaucrati positions as the Earl rose to become Chancellor.

Locke was interested in philosophy, and it was the writings o Descartes in particular which first interested him. As Locke put it: he wante to understand very precisely and systematically what knowledge "was capabl of."[83] Nevertheless Locke was too involved with the vagaries of Britis politics to write early in his life. In 1683 he was even forced to slip away int exile in Holland following the Rye House Plot to kidnap the King. Locke wa able to return to Britain in 1689 following the crowning of William o Orange, and it was at this time that the majority of his works were finally printed.[84]

The Essay Concerning Human Understanding, (1690) his magnum opus on epistemology, was inspired by a conversation with a group of friends in 1671. They were engaged in philosophical discourse, when it became clear that they could make no further progress until they had examined the mind's capacities and had seen "what objects our understandings were or were not fitted to deal with."[85]

---

[83] Dunn, John Locke (Oxford: Oxford University Press, 1984) 7
[84] Locke actually wrote a number of his more famous treatises while in exile.
[85] Copleston, Frederick A History of Philosophy Vol.5 (Westminster: The Newman Press, 1964) 67

Locke's basic notion counters Descartes in that he believes that xperience is the basis for all knowledge. We receive "ideas" from sense xperience, and Knowledge, with a capital "K", is the perception of the greement or disagreement of two ideas. There are four means of establishing nowledge: Identity, Relation, Co-existence or Necessary Connection and Real Existence. All knowledge is also either actual (directly in front of us) or abitual (having seen proof and remembering it.)

Locke's theory of knowledge, labeled Empiricism, is therefore based n trusting one's senses and intaking experience, and registering them as imple ideas, or collecting them into complex ideas. Therefore there is a ubjective nature to Locke's theory, which he addresses. How can we know, a potential objector asks in Book Four, whether knowledge is real − 'knowledge placed in our ideas may be all unreal or chimerical." To this Locke replies that ideas must agree "with things."[86] He goes on to say that "our knowledge therefore is real only so far as there is conformity between our ideas and the reality of things."[87] This is a far cry from Descartes' certainty. And yet he still attacks Descartes later in that chapter arguing "How vain I say, it is to expect demonstration and certainty in things not capable of it... and refuse assent to very rational propositions, and act contrary to very plain and clear truths, because they cannot be made out so evident as to surmount every the least (I will not say reason, but) pretense of doubting." How scathing is this − to argue that this is not reason − that Descartes' doubt

[86] Locke, John An Essay Concerning Human Undersanding in Great Books of the Western World Robert Maynard Hutchins, Editor (Chicago: Encyclopedia Brittanica, 1952) 323
[87] Locke, 324

is not reason! And this is the true advance which Locke brings to Wester epistemology. For Locke, if we insist on certainty we lose all our bearings.[88]

One might be tempted to believe, then, that Empiricism places the bod in a positive liminal space because it offers a faith in the senses. But in fac Locke's vision of the body, or belief in the body as a source of knowledge does not change the direction of the previous hundred years' developments i epistemology. As Morris Berman notes, "there was no real clash betwee Rationalism and Empiricism. The former says that the laws of though conform to the laws of things; the latter says always check your thought against the data so that you know what to think."[89] Locke does not see th body itself as a source of knowledge, or challenge Descartes' thinking o emotion as being placed in the body. Locke merely sees the body's facultie as a tool, a machine that can be trusted empirically to provide input. Therefor Locke's Empiricism is not a step toward empathic knowledge, but in fac another step toward the mechanization of knowledge. The body is ar instrument to produce metered stimuli, no more.

As Copleston writes, Locke was actually also a Rationalist in that "he believed in bringing all opinions and beliefs before the tribunal of reason and disliked the substitution of expressions of emotion and feeling for rationally grounded judgments."[90] Locke's rational judgments included the senses, and did not extend to an embodied re-assessment that would place power in embodied knowledge such as emotion or feeling. Locke's major contribution to the field was very simple. He saw the body as an instrument, a tool or vessel that provided reliable input. Locke's Empiricism may have conflict

---

[88] Aarsleff, Hans "Locke's Influence" in The Cambridge Companion to Locke Ed. by Vere Chappell (Cambridge: Cambridge University Press, 1994) 264
[89] Berman, Morris The Reenchantment of the World (Ithaca: Cornell University Press, 1981) 28
[90] Copleston, 1964, 69

with Descartes in certain ways, but in the manner that the body was deeply understood, there is more harmony than discord. [91] And though Locke's Empiricism replaced Descartes' radical doubt with a more reasonable faith in the ability of the senses, the overarching reductionism and control of reason still presided. The result was to maintain a strong body/mind dichotomy.

The successful use of the mechanistic, mathematical method, cemented a human conception separate from the body. Simultaneous was a full division of the human from the natural world. The commonality that we had with the natural world – our nature, our body – was no longer critically human. Knowledge generation became broadly accepted as something of the mind alone. Emotion, which was already associated with passion, and frought with spiritual difficulty, was cemented to the body. The following chapter will address three movement training techniques, each of which addresses, in its own way, the possibility of overturning the hardened body/mind duality. Simultaneously, the function of opening to the body as a source of knowledge is recognized.

---

[91] From the very beginning of the Essay, Locke is clearly of a mind to set the record straight regarding his opinion of Descartes. Just after introducing his method – that he will search out the boundary between opinion and knowledge and the means to know the two – Locke offers the following rebuke of Cartesian doubt. "It will be no excuse to an idle and untoward servant, who would not attend his business by candle light, to plead that he had not broad sunshine. The candle that is set up in us shines bright enough for all our purposes....If we will disbelieve everything, because we cannot certainly know all things, we shall do much what as wisely as he who would not use his legs, but sit still and perish, because he had no wings to fly." In Locke's vision, our capacities and abilities are given us by God and we should accept our limitations. Locke also attacks Descartes' dream proof of the separateness of body of mind, arguing in Book Two Chapter 16 that "if its separate thoughts (while asleep) be less rational, then these men must say, that the soul owes its perfection or rational thinking to the body; if it does not, it is a wonder that our dreams should be, for the most part, so frivolous and irrational; and that the soul should retain none of its more rational soliloquies and meditations." Further, while he agrees with Descartes on issues such as free will he directly attacks the notion of an independent mind and body, arguing in Book Two Chapter 27 Verse 15 that "the body, as well as the soul, goes to the making of a man."

Chapter 4: Somatic Knowledge Development - Contact Improvisation, Skinner Release Technique and Body-Mind Centering

"As long as my body talks, I can't help but listen"

-Helen Clarke[92]

Thomas Hanna first described the term Somatics in 1976, defining it as: "The art and science of the inner-related process between awareness, biological function and environment; all three factors are being understood as a synergistic whole."[93] Dr. Hanna chose the ancient Greek word Soma, meaning 'the living body', to designate this area of study.

The term Somatics has come to include not only fields of inquiry (including physical therapy, and dance) but an understanding of the body from within. The term Somatics is increasingly used to encompass systems utilized to create a clearer relationship between the mind and body. In addition to inspiring the theory of Somatic Ecology, existing Somatic study enables the central conclusions of Somatic Ecology to be achieved.

Somatics is a recognized arena within the existing scientific/cultural framework that validates and develops Somatic Knowledge. As Dance

---

[92] Contact Quarterly Vol. 3 #2 The Long Winter 1977/78 pg.4
[93] Hanna, Thomas cited on www.Somaticsed.com within *What is Somatics* by Dr. Eleanor Criswell Hanna, Ed.D.

professor Shelly Bickels noted, "Somatic studies do not negate scientific third-person objective examination of the body....[Dr. Hanna wrote in 198 that] there is an emphasis in Somatics on 'the complementary relationship between science and Somatics. Together, the two viewpoints create a authentic science that recognizes the whole human being.'"[94]

As explored in chapter 2, and as will be explored more completely i chapter 5, validation of Somatic Knowledge is critical to overcoming th humanity/nature divide at the root of the Deep Ecology problematic. Prio chapters have detailed the dissociation of the human body from the spirit an mind. A product of that dissociation is the elimination of Somatic Knowledge or more accurately, a cessation of the validation of Somatic Knowledge. The prior chapter noted several stages in the development of the scientific method and resulting separation of the human from the rest of the natural world. By the beginning of the 20th century, the human body – which for millenia was a central tool for human understanding- was devalued, and debased. Somatic study is available to reverse this relatively recent human re-valuation.

Within the modern years of human/nature divorce, and mind/body divorce, there have regularly risen divergent philosophies that address the concerns of mind/body divorce. From the philosophy of Husserl and Merleau-Ponty, to the new age/new philosophy of the Hippie Generation, regular backlash against the silencing of empathic Knowledge has emerged. This chapter does not consider modern arguments for Somatic Knowledge within the field of epistemology, but rather focuses on Somatic Knowledge Development systems consistent with the existing theoretical framework.

---

[94] Bickels, Shelley MA Thesis American University, 1998 (32-33)

59

Somatic Knowledge Development systems have existed within numerous physical, academic, cultural, and spiritual practices. Since just the middle of this past century, in the United States and abroad, scores of new Somatic Knowledge Development methods have emerged. A small number of these new Somatic Knowledge Development systems have been codified and documented sufficiently to warrant academic curricula and research. This chapter will consider three such systems, each currently taught broadly in U.S. Colleges and Universities. The study of Contact Improvisation (CI), Skinner Release Technique (SRT), and Body-Mind Centering (BMC) occur largely within Dance Departments. The study of Somatics and Somatic Knowledge Development systems is not limited to the field dance. This book considers dance-focused Somatic practices as a necessary means to narrow the breadth of inquiry toward proving the viability of the theory of Somatic Ecology. More comprehensive cross-disciplinary research defining linkages between Somatic Knowledge Development and various non-arts-related professions would be worthwhile.

Current work in Somatics does not impose rigid technique on the body, but provides a variety of frameworks/practices/mechanisms, encouraging the body toward awareness. Contact Improvisation (CI), Skinner Release Technique (SRT), and Body-Mind Centering (BMC) are in use today within the field of dance and movement re-education to develop Somatic understanding. As techniques they share commonalities with other forms of somatic training, including martial arts, ballet, etc. What distinguishes these techniques for the purpose of this investigation is their validation of the existing body in time and space. CI, SRT, and BMC place value in and utilize the body as a source of Knowledge. As they relate to Somatic Ecology, they

60

are examples of viable existing means to challenge the modern body/mind duality, within a modern scientific theoretical framework.

Contact Improvisation (CI), was initially developed by Steve Paxton, a dancer and choreographer. The term was coined in 1972 to describe a movement form that combined physical and mental skills.[95] The movement problems and exercises developed in CI training are based on sensing the use of weight and force in one's body, and working with other bodies to allow challenging, subtle, unexpected and unusual relationships to develop. Some of these relationships involve complete transference of weight – carrying and lifting – and subsequent release, sometimes involving falling. The unusual movements and relationships that occur in CI classes, jams, and performances are made possible by listening to the body, as opposed to planning with the mind. CI training emphasizes that the messages received by a body in a state of awareness are necessarily conjoined with an equally active, aware mind. Practitioners strive to attain a heightened sense of consciousness in which body and mind function in concert.

Steve Paxton explained the basic premise of CI in a 1979 article:

> The exigencies of the form dictate a mode of movement which is relaxed, constantly aware and prepared, and on-flowing. As a basic focus, the dancers remain in physical touch, mutually supportive and innovative, meditating upon the physical laws

---

[95] Some consider the beginning of CI to have come with the performance of a piece called "Magnesium" during a residency conducted by Paxton in the Winter of 1972 at Oberlin College. Others look to a group of athletes and dancers that Paxton gathered together for a week of five hour a day performances at the John Weber Gallery in New York City in July 1972. But even before those first performances of "Contact" Paxton had been working on the principles, and he, and others, have continued to refine the approach.

relating to their masses: gravity, momentum, inertia and friction. They do not strive to achieve results but rather, to meet the constantly changing physical reality with appropriate placement and energy.[96]

CI requires both the mind and the body to be constantly open and prepared, and the grounding for this started in the initial classes. Daniel Lepkoff, one of the original contact performer/teachers who is still active today, describes some of Paxton's early focus on the body from a workshop at the University of Rochester in 1971.

There was lots of time, time to listen, to Steve's thoughts and images, and time to sit in silence, being aware of our physical sensations and perceptions, conscious of 'one's animal' (a concept Steve often used in these early years)... 'one's animal' is a physical intelligence composed of movement patterns and reflexes, both inherited and learned, that form our ability to survive and meet and play energetically with our environment. A main aspect of the early Contact Improvisation work sessions was to coax, encourage and engage this animal intelligence. Because 'one's animal' was understood to be a deep aspect of the physical self, much of our efforts were put towards getting out of its way, letting go of one level of control and learning to trust in another.[97]

---

[96] Paxton, Steve Contact Quarterly Vol. 4 #2 Winter '79 pg. 26
[97] Lepkoff, Daniel, Contact Quarterly Winter Spring 2000 pg. 62

CI engages a physical intelligence, sacrificing to some extent the rational faculty that tends to dominate every day life. In an article published in Contact Quarterly in 1982, Paxton described dancing CI, saying, "One thing is clear. I have little memory, muscular or mental, of what I've danced. The specific movements my body executes when I improvise do not register consciously, and I can't reconstitute them. I feel transparent in the action. Causing it only a little, and holding no residuals."[98] The sense of complete momentary understanding, without concern for the understanding's documentation, is a hallmark of Somatic Knowledge. It exists within the lived moment. This type of Knowledge can also be understood as empathic understanding, or sympathy with life (as Deep Ecologist's describe it.)

Learning to engage the mind with Somatic sensitivity is central to the development of the CI skill set. An example of the CI skill set's relationship to the development of Somatic sensitivity can be viewed in one of the seminal exercises developed by Paxton, "the small dance." In describing "the small dance" Paxton stated, "we're trying to get in touch with these kinds of primal forces in the body and make them readily apparent. Call it the small dance... and your mind is not figuring anything out and not searching for any answers or being used as an active instrument but is being used as a lens to focus on certain perceptions."[99] If one considers the body the focus in Contact Improvisation, the mind is the lens that brings it into focus. And if one considers the mind the focus, it is the body that serves as the lens. Somatic Knowledge is validated.

---

[98] Paxton, Steve <u>Contact Quarterly</u> Vol. 7 #3/4 Spring/Summer 1982 pg. 17
[99] Paxton, Steve interviewed by Elizabeth Zimmer on CBC Radio <u>Contact Quarterly</u> Fall 1977 Vol. 3 #1 pg. 11

CI is a dance technique that requires a certain set of physical skills but also, more significantly, requires a transparent body/mind state. In the process of practicing the technique it is clear that being attuned with one's body is as important as being in touch with one's dancing partner. The methodology, and the practice, of getting in touch with "your animal" is an integral part of the technique. As Contact practitoner/teacher Jerry Zientara argues, "In Contact Improvisation what the learner learns is her/his own body, going to the body as the source of knowledge and delivering to the body all the information gathered."[100] As with other forms of Somatic practice, Contact Improvisation focuses on inner and outer awareness. The goal is not to learn a specific set of movements, but to practice a broader understanding of Knowledge in the body.

Contact Improvisation is not the only modern technique keyed on validating and developing Somatic understanding. In a 1979 article by Joan Skinner, Bridget Davis, Sally Metcalf and Kris Wheeler, the authors wrote that, "[Skinner Release Technique] SRT can be described as a system of kinesthetic training which refines the perception and performance of movement through the use of imagery."[101] In teaching the technique, students are instructed to first become more attuned to their senses and how they

---

[100] Zientara, Jerry "A Few (Tight) Marbles From My Teacher's Notebook" Contact Quarterly Spring/Summer 1978 Vol. 3 No. 3/4 pg. 10
[101] Ibid, pg.8

move, and secondarily to be influenced creatively and thoroughly by the images.[102]

The authors go on to assert that "in SRT the image serves as the carrier of a patterned whole of information – a metaphor for kinesthetic knowledge – which 'formulates a new conception for our direct imaginative grasp,' and this metaphor is apprehended intuitively rather than analytically."[103] The SRT metaphors work on multiple levels, physical and poetic, and ask the body/mind to engage creatively. As with CI, the mind are body are equal partners in Knowledge development. The concept that the work is intuitive rather than analytical is significant. Skinner seeks to subvert the mechanistic rational learning process in favor of non-linear, sub-rational learning process. Skinner discusses the process saying:

> The kind of learning we're tapping into, which is not A leading to B leading to C, is the way I think we actually learn…The images we use are not anatomical; they're more indirect and oblique – deliberately so because we're getting at energy levels that are even underneath the anatomy of the body. They have several kinds of purposes. Some encourage stillness and just being there; some stimulate movement; some stimulate this letting-go process, which is letting go of so many things, not just physical holding patterns, locks, but also preconceptions and expectations.[104]

---

[102] It should be noted that Skinner is only one of a set of "release techniques" in the pantheon today. There are other versions that either are or have been taught by: Mary Fulkerson, Nancy Topf, Susan Klein and Barbara Mahler.
[103] Ibid, pg.8
[104] Skinner, Joan interviewed by Agis, Gaby in <u>Dance Theater Journal</u> pg. 37

Skinner engages the mind on the creative level so that SRT can get beyond the established pathways that exist in the body/mind.[105] As she said in a 1990 interview:

> Something compelled me to avoid the analytical work while working with this. It seemed as if they were functioning in two different realms of knowing and of experiencing. I don't know if I can answer why I chose the poetic route, but it just seemed absolutely essential.... I avoid the direct one-to-one muscular approach – not that it's invalid in any way, just that it gets in the way of going through *this* door to experiencing.[106]

SRT involves a released creative process, similar to the CI process as described by Lepkoff. SRT training asks students to move beyond analytical control, and to allow the body as a medium, in the sense of a conduit for exploration and discovery. The technique uncovers and connects body and mind through this process. As choreographer/teacher/interviewer Stephanie Skura noted lauding the technique, "one thing that disturbed me when I was studying dance was that nobody was talking or teaching about what was going on inside you. (In Skinner) there was an acknowledgement of the relation of mind and body."[107] Through poetic image work SRT guides the practitioner into a new relationship with the body, one which significantly re-

---

[105] This is the root of the name for the form. As Joan Skinner describes, "it was the students who coined the word releasing. I guess I must have been saying, 'we're releasing this and that... we're releasing the breath and the tension pattern. So they called it the releasing technique and it's been evolving ever since that one year." (Skinner, Joan, interview by Agis, Gaby in <u>Dance Theater</u> <u>Journal</u> pg. 35)

[106] Skura, Stephanie "Releasing Dance: Interview with Joan Skinner" <u>Contact Quarterly</u> Fall 1990, pg. 12

[107] Ibid, pg. 16

introduces the dancer to their body. This process of child-like creative exploration profoundly validates embodied knowledge, and the partnership between body and mind.

SRT exists within the field of Somatics in its contemplative and holistic use of the whole self - body and mind. SRT engages the individual with the whole environment, inner and outer. SRT asks that old patterns be released so that clearer, cleaner, and healthier ways of moving can be discovered. Similar to the concerns of environmentalists, this requires a renewed relationship with the body as a source for Knowledge.

Body-Mind Centering (BMC), a system originated by Bonnie Bainbridge Cohen, shares SRT's understanding of release and re-discovery as a core training function.[108] The system is consistent with CI and SRT's development of mind/body dialogue. A long-time student/teacher Linda Hartley notes, "In the Body-Mind Centering approach we recognize that body and mind have distinct functions; experiencing the body from within, we come to see that they are integrally connected aspects of a greater whole."[109] Hartley describes the components of BMC as follows:

[The training] involves direct experience of anatomical body systems and developmental movement patterns, using techniques of touch and movement repatterning; central to the work is the

---

[108] Originally trained as an occupational therapist, between 1962 and 1972, Ms. Cohen worked with severely disabled adults and children. She was subsequently certified as a Neuro-development Therapist by the Bobaths in England, and studied Neuro-muscular Re-education with Andre Bernard, Katsugen Endo (the art of training the nervous system) with Haruchi Noguchi, and dance therapy with Marion Chase. In 1973 she founded the School for Body/Mind Centering.

[109] Hartley, Linda Wisdom of the Body: An Introduction to Body-Mind Centering (Berkeley: North Atlantic Books, 1995), xxv-xxvi

process of awakening awareness at the cellular level to contact the innate intelligence of the body; in BMC practice (the development process) is explored and embodied through a series of movement patterns... (and) movement is re-patterned as we allow the 'mind' of our learned movement patterns to change; Body-Mind Centering also involves an in depth and experiential study of all the anatomical systems of the body. The musculoskeletal system and the organs, glands, nervous, and fluid systems each express their own quality of movement, feeling, touch, perception, and attention.[110]

BMC involves engaging the self, in the Somatics senses of Sensorium and Motorium, to touch cellular awareness, developmental process, and the anatomical systems of the body. Like SRT and CI, the exercises develop mind and body together, increasing the range of each.

There is a visionary component to BMC that is developed through tuning in to each 'aspect' of the body. Each process and system is used, while in a meditative clear body/mind state to discover movement. The movements might be called improvisations were not the intent and focus quite so concrete. The movement itself is analyzed as a type of output, and is treated within the BMC practice as Knowledge.

Where SRT engages creativity to find the body, BMC teachers engage in scientific instruction to encourage a more or less creative expression by the body of that understanding. BMC and SRT both deeply validate the body's

---

[110] Ibid, xxix-xxxii

68

understanding. BMC provides a new language to articulate the Knowledge gleaned from the body. Hartley concludes:

> (BMC) offers a language for what we already know through the innate wisdom of the body, and for what we express, consciously or unconsciously, through our actions. By defining our experiences through language in this way, these experiences may be remembered, brought to consciousness, clarified, seen, or articulated in a new way. The act of bringing to consciousness and into language that which was unconscious, unknown, is in itself a process of empowerment.[111]

A centerpiece of the new language is a realignment of the body/mind stasis. This occurs through a reconstitution of "mind." The mind is not articulated as a group of cells in the skull, alone. In BMC, each part of the body, each system, has its own mind/understanding. As Hartley describes,

> [Body-Mind Centering language speaks of the] "mind of a particular body system: skeletal, muscular, organ, etc.... For the purposes of my description of Body-Mind Centering I will use the term "Mind" in quotes when describing a particular body-mind experience. A specific "mind" can be experienced and witnessed when we direct our attention to a particular body system or part of the body, or when we move with a certain focus and identifiable quality. What we experience and observe is a

---

[111] Ibid, 306-7

particular quality of awareness, feeling, perception, and attention when we embody a movement pattern or body system; this is the "mind" of that pattern or system, and is an expression of the integrated body-mind.[112]

BMC technique integrates a vocabulary to integrate its more expansive understanding of Knowledge, and the inclusion of Somatic Knowledge. "Mind" is not simply cognitive function or the controller of the body, but is an expression of awareness/understanding within each body. The language of BMC reflects integration of the body as a non-subjugated subject to the mind as a non-controlling partner.

Hartley states that "the heart of this… contains a search for the wisdom that we all possess within us, the awareness of who we are on this earth. By journeying deep inside to our own experience and out through our perceptions to the world we live in, we can begin to see who may truly be, beyond conditioned self-images and habitual patterns of thinking, moving and living. The medium for our research will be the body and its movement."[113] Or as Bainbridge Cohen states, "I am interested in the process of the unfolding of the self."[114]

BMC shares a strong sensibility of training/re-training the whole human. At the core is a new balance in the valuation of the human body and human mind. As in Deep Ecology, the new balance is pursued through re-discovery of the self in the world. CI, SRT, and BMC go to the body in search

---

[112] Ibid, xxv-xxvi
[113] Hartley, xix
[114] Bainbridge Cohen, Bonnie in "Interview with Bonnie Bainbridge Cohen" Nelson, Lisa and Stark Smith, Nancy Contact Quarterly Winter 1980 pg. 28

of that self, in search of the knowledge that will enrich the lives of it practitioners.

It has occurred to practitioners of these Somatic Knowledg Development systems that changing the body/mind relationship, and th relationship to the self, has implications for the relationship to the planet. I Hartley's manual on BMC, several passages mirror texts by Deep Ecologists Hartley states,

> "This story is about embodiment, the human being at home, each of us in our own body. To be present in our body is a form of awareness, and it is a first step toward being kind to ourselves and others. In coming into our body we become connected to our greater home, the earth; we become a part of the earth and she a part of us…..Too often, perhaps especially in modern Western culture, the union of body and mind, of the "earth" and "heaven" principles, is not harmonious. One is often overemphasized at the expense of the other, or one aspect may be denied, causing the other to suffer from exhaustion and distortion."[115]

CI, SRT, and BMC provide practitioners with a way of knowing that is simultaneously intimate and scientific. The Somatic Knowledge Development process opens the door to a new relationship to ones inner and outer environments.

The field of Somatics, inclusive of these three techniques, has a strong focus on developing consciousness and awareness. As Bickels notes,

---

[115] Hartley, 1995, xxi

Consciousness designates the range of voluntary sensory-motor functions acquired through learning, which occurs from birth onward.... Awareness is the ability to isolate and choose specific activity and focus on it. Thus a somatic learning process involves focusing awareness on something unknown, trying to associate it with something already known. The goal is to cause the unlearned to become learned, and the forgotten to become re-learned."[116] Somatic techniques validate embodied Knowledge, and a present dialogue between the body and mind.

Contact Improvisation, Skinner Release Technique, and Body-Mind Centering are three available tools to reconfigure the understanding of the Human in the world, through a new understanding of the Human in the body. These three techniques prove that there are available, enjoyable, means of validating Somatic Knowledge. Somatics' emphasis on the awareness of inner and outer environments is consistent in these three methodologies, which while stimulating body knowledge do not sacrifice connection to the greater world. Somatics and Ecology both develop from sensitivity, to empathy. Somatics and Ecology both then proceed from empathy to sympathy, leading to a new harmony of the human in the world. The following chapter will analyze the connections between the issues raised by environmentalists with the solutions that Somatic Knowledge Development systems offer.

---

[116] Bickels, 1996, 29

Chapter 5: Somatic Ecology

"To know your own flesh, to know both the pain and joy it contains, is to know something much larger than this."[117]

– Morris Berman

This chapter places Somatic Ecology in context, with examples for the theory's application and relevance. The prior chapters have shown that the Humanity/Nature divide cannot be addressed without addressing the Human/Human body division. The first chapter of this book introduced the Environmental Historian David Pepper arguing, "Ecology conveys the universal principles about how humans ought to behave in nature." [118] One comes to understand from Deep Ecologists that it is the conception of Humans as *separate* from nature that is to blame for the environmental crisis.

As explored in Chapter One, Deep Ecology argues that in order for a full balance in the intellectual, emotional, and physical realm, human beings need to sense their part within that whole. The Human as part of nature, inside nature, embodied in nature, is necessary. The study of Somatic practices in Chapter Four validates the role that Somatic training can play in accomplishing that goal.

In chapter 4 three Somatic Knowledge Development mechanisms were explored. One in particular developed not only a new sensibility of the Human in Nature, but also a new language to describe the phenomenon.

---

[117] Berman, Morris Coming to Our Senses: Body and Spirit in the Hidden History of the West (NY: Simon and Schuster, 1989) 344
[118] Pepper, David Modern Environmentalism pg. 240

Body-Mind Centering invites practitioners to a new understanding of the word 'Mind'. Each of the three techniques examined developed Somatic Knowledge, and noted that in so doing the Human relationship in the world is re-valued.

The modern human conception was built upon non-rational human understanding. While the science of Ecology may lead to prescription regarding human behavior, that behavior is discouraged as long as the Human conception – separate from its own nature – is maintained. The dominant relationship of man toward the natural world is sanctioned by the idealogy that sets human in opposition to nature. This is the same ideology, which even in prescription, declared that the issue is how 'humans behave in nature'. A revaluation of the Human conception is possible through Somatic Knowledge Development.

The body is that part of nature which we are most intimate with. It is that part of nature which is human, which is *us*. The Human relationship to the body mirrors our relationship to nature. It is beyond metaphor, it is actually reflective of the exact relationship that this culture has toward the natural. Transforming the link to our natural selves, which would occur through a rejection of historical relationships of domination and of historical dualisms, would generate vastly different inner and outer worlds.

Deep Ecology and Somatics both offer arguments for an end to dualities. Somatic Ecology argues that the lessons which dance/somatics offer – trusting the body, listening to the integrated body/mind, being attuned to the sensual world - are a means to enter into a new way of knowing, which, it has been argued by Deep Ecologists, is truly necessary. Reconnecting the human with nature, through Somatic Training or other means, requires becoming tuned to embodied knowledge. Embodied knowledge is that type of knowledge that we share with all living entities. The word that describes this type of feeling is "empathy." The world of empathic knowledge is filled with real, but subjective, commonalities.

Asserting the worth of embodied Knowledge is a slippery slope. The validation of Somatic knowledge is not in any way anti-intellectual. It is rather, an assertion of multi-intellectuality. Validating embodied knowledge involves valuing both objective and subjective knowledge. The editors of _Women's Ways of Knowing_ argued that:

> The shift into subjectivism is, we believe, a particularly significant shift for women when and if it occurs.... as she begins to listen to the 'still small voice' within her, she finds an inner source of strength. A major developmental transition follows that has repercussions in her relationships, self-concept and self-esteem, morality, and behavior. Women's growing reliance on their intuitive processes is, we believe, an important adaptive move in the service of self-protection, self-assertion, and self-definition. Women become their own authorities... In a world that emphasizes rationalism and scientific thought, there are bound to be personal and social costs of a subjectivist epistemology.[119]

Subjectivism is a tool for self-definition, and empowerment. The cultural historian Morris Berman argues that objectivity works against true

---

[119] Belenky, Mary Field; Clinchy, Blythe McVicker; Goldburger, Nancy Rule and Tarule, Jill Mattuck _Women's Ways of Knowing_ (New York: Basic Books, 1986) 54-55

understanding, and notes that our current stance is truly a cultural artifact "prior to 1600 *lack* of identification was regarded as strange."[120] Writing in Mind and Nature, Gregory Bateson similarly noted, "epistemology is always and inevitably personal." [121] Validating subjective Knowledge does fly in the face of the Androcentric worldview. Such a revaluation is long overdue.

The scientific method is overwhelming positive. Somatic Knowledge Development systems do not invalidate science. In Chapter 4 Daniel Lepkpof was quoted saying that Somatic training involves, "letting go of one level of control and learning to trust in another." The environmental crisis provides the reasoning for validating embodied knowledge. Somatic practices provide a methodology for pursuing that end.

How we think about things matters. Our environmental problems increase every day, because the guide to our actions – our thought – is removed from empathy with the natural world. We move forward simply because we can, because we feel that it is our destiny to control, to dominate.[122] As Gregory Bateson put it, "if we have wrong ideas of how our abstractions are built – if in a word we have poor epistemological habits – we shall be in trouble – and we are."[123]

Deep Ecology seeks to end the anthropocentric worldview, and bring about a paradigm of self-realization. Ecofeminists seek to connect an end of historical dualisms to an end to patriarchy. These two ethics are easily related to Somatic Ecology, as both have at their heart a rejection of dualisms, an

---

[120] Berman, Morris Coming to Our Senses: Body and Spirit in the Hidden History of the West (New York: Simon and Schuster, 1989) 111-112
[121] Bateson, Gregory Mind and Nature: A Necessary Unity (New York: Bantam Books, 1979) 8
[122] Bill McKibben addresses this loss in his elegant book Enough. McKibben, Bill Enough : Staying Human in an Engineered Age (New York: Times Books, 2003)
[123] Bateson, Gregory A Sacred Unity: Further Steps to An Ecology of Mind ed. by Rodney E. Donaldson (New York: Harper Collins, 1991) 233

overthrow of domination, and a reassessment of our ingrained beliefs. They both support a premium being placed on empathy, and the connection to the natural world. They both support a reinvigoration of our natural selves; in Deep Ecology as a result of the process of self-realization, and in Ecofeminism as a part of the process of overthrowing the patriarchal influence of historical dualisms.

Ecofeminism and Deep Ecology both make arguments about the relationship of domination and dualism, without overtly reflecting on the consequences. The body is a significant source of Knowledge. Conceptualizing the body this way, Somatic theorists and practitioners counter the paradigm that the mind is the only reasonable means for understanding the world. They validate empathy with the natural world, and sympathy with all life. The basic underlying premise of much work in Somatics is that the mind and the body must have constant interchange and exchange, and that they are equally important for our healthy and knowledgeable existence in this world.

In a world where more and more people live in cities it has to be a priority to find ways to allow people to connect with nature without necessarily having to take a walk in the woods. Somatic training can serve toward that end. Dance can be used to teach environmentalism as it teaches young – or old – dancers to become attuned to the natural world: their personal natural world. We can, in dance classes, rediscover our wild selves and so reconnect to the wild and natural around us, even in an urban environment.

There are several predictable outcomes from the validation of Somatic Knowledge. Today, Deep Ecologists are forced to make arguments for their new ethic based on rational, objective knowledge. That is the only type of knowledge accepted in the academic, literate, scientific community. Validating empathic knowledge, Deep Ecologists would be able to argue for the things that are most important to them, using the tools that are most germane to the job.

While the science of Ecology offers prescription for how Humans ought to behave in nature, what truly motivates most environmentalists is not statistics, but an embodied empathy with the world, a sense of belonging and trying to live a just life. Writers from Aldo Leopold[124] to Henry David Thoreau[125] to Edward Abbey[126] have noted this in their writings. In validating subjective/empathic understanding, Somatic Ecology might support new arguments around the prioritization of resources.

Researching the philosophers of the sixteenth to the eighteenth century – their lives and their methodologies – a common commitment to a greater whole was visible. As science historian Paulo Rossi writes in <u>The Birth of Modern Science</u>,

> The leaders of the Scientific Revolution believed that re-establishing human control over nature and the advancement of learning were of value insofar as they were part of a larger context that involved religion, morality, and politics. Campanella's 'universal theocracy', Bacon's 'charity', Leibniz's 'universal christianity', and Comenius's 'universal peace' cannot be divorced from the interest and enthusiasm of

---

[124] Leopold, Aldo <u>A Sand Couny Almanac and Sketches Here and There</u> (New York: Oxford University Press, 1968, c.1949)
[125] Thoreau, Henry David <u>Walden: or Life in the Woods</u> (New York: C.N Potter, c.1970)
[126] Abbey, Edward <u>Desert Solitaire: A Season in the Wilderness</u> (New York: Simon and Schuster, 1968)

these men from the new science. Nature was at the same time an object of domination and of reverence. It was to be 'tortured' and bent to serve man, but it was also 'God's Book' to be read with humility.[127]

In centuries past there was a purpose to existence that centered in one's relationship to an ineffable deity. Knowledge was not an end unto itself. Today most people have lost that purpose. It is interesting to consider (as addressed in Chapter Two) that rediscovering a relationship to one's animal self, to one's soma, can help one find empathic connections with other life. In the same way that native peoples find meaning through interactions with the larger world, those connections may provide context to our new modern lives.

As humans grapple with the global environmental crisis we cannot count on a single faith to unite our efforts. Our commonality exists only in our fully embodied humanity. Sustainability, if not morality, is certainly an adequate measure of a culture's sanity. The exigencies of earlier environmental theories support investment in Somatic Training as a solution to the root of the environmental crisis. Bridging the Humanity/Nature divide through validation of embodied knowledge may result in a healthier global and human ecology.

---

[127] Rossi, Paolo The Birth of Modern Science Translated by Cynthia de Nardi Ipsen (Blackwell: Oxford University Press, 2000) 40

Bibliography

Aarsleff, Hans "Locke's Influence" in <u>The Cambridge Companion to Locke</u> Ed. by Vere Chappell (Cambridge: Cambridge University Press, 1994)

Abbey, Edward <u>Desert Solitaire: A Season in the Wilderness</u> (New York: Simon and Schuster, 1968)

Baker, Cynthia M. "Ordering the House: On the Domestication of Jewish Bodies" in <u>Parchments of Gender: Deciphering the Bodies of Antiquity</u> ed. by Maria Wyke (Oxford: Clarendon Press, 1998)

Bateson, Gregory <u>Steps to an Ecology of</u> Mind (New York: Ballantine Books, 1972)

_____ , <u>Mind and Nature: A Necessary Unity</u> (New York: Bantam Books, 1979)

_____ , <u>A Sacred Unity: Further Steps to An Ecology of Mind</u> ed. by Rodney E. Donaldson (New York: Harper Collins, 1991)

Belenky, Mary Field; Clinchy, Blythe McVicker; Goldburger, Nancy Rule and Tarule, Jill Mattuck <u>Women's Ways of Knowing</u> (New York: Basic Books, 1986)

Berman, Morris The Reenchantment of the World (Ithaca: Corne
University Press, 1981)

_____ , Coming to Our Senses: Body and Spirit in the Hidde
History of the West (NY: Simon and Schuster, 1989)

Berry, Wendell Home Economics (San Francisco: North Point Press
1987)

Boorstin, Daniel J. The Seekers: The Story of Man's Continuing Ques
to Understand His World (New York: Random House, 1998)

Brower, David with Steve Chapple Let the Mountains Talk, Let the
Rivers Run: A Call to Those Who Would Save the Earth (San Fransisco:
Harper Collins West, 1995)

Brown, Peter The Body and Society: Men, Women and Sexual
Renunciation in Early Christianity (New York: Columbia University Press,
1988)

Clarke, Helen Contact Quarterly Vol. 3 #2 The Long Winter 1977/78
pg.4

Copleston, Frederick A History of Philosophy Vol. 4 (Westminster:
The Newman Press, 1961)

_____, A History of Philosophy Vol. 5 (Westminster: The Newman Press, 1964)

Descartes, Rene Discourse on Method, (Indianapolis: Hacket Publishing Company, 1998, 1637)

_____, Meditations on First Philosophy (Indianapolis: Hackett Publishing Co., 1641; 1998)

Drake, Stillman Galileo (New York: Hill and Wang, 1980)

Greenberg, Blu How To Run a Jewish Household (Northvale: Jason Aronson, 1989)

Dunn, John Locke (Oxford: Oxford University Press, 1984)

Fermi, Laura and Bernardini, Gilberto Galileo and the Scientific Revolution (New York: Basic Books, 1961)

Fulkerson, Mary "The Release Class" Movement Research Performance Journal #18 Spring 1999

Gere, David "Talking Dance 11/16/87 – Introducing Anna Halprin" from Personal Papers of Anna Halprin, NYLPA MGRL 98-4751

Halprin, Anna Lomi School Bulletin (Mill Valley, Ca) Summer 1981

Halprin, Anna ed. by Goldsmith, Carolyn "The Child as Creator" Personal Papers of Anna Halprin, NYLPA MGRL 98-4751

Halprin, Anna <u>Moving Towards Life</u> (Hanover: Wesleyan University Press, 1995)

Hartley, Linda <u>Wisdom of the Body: An Introduction to Body-Mind Centering</u> (Berkeley: North Atlantic Books, 1995)

Kerner, Mary C. "Anna Halprin: Integrating Emotion and Technique" <u>Dance Teacher Now</u> June 1988

Kheel, Marti "Ecofeminism and Deep Ecology: Reflections on Identity and Difference in <u>Reweaving the World: The Emergence of Ecofeminism</u> Irene Diamond and Gloria Feman Orenstein Eds. (San Fransisco: Sierra Club Books, 1990)

King, Ynestra "Healing the Wounds: Feminism, Ecology, and the Nature/Culture Dualism" in <u>Reweaving the World: The Emergence of Ecofeminism</u> Irene Diamond and Gloria Feman Orenstein Eds. (San Fransisco: Sierra Club Books, 1990)

Lahar, Stephanie "Roots: Rejoining Natural and Social History" in <u>Ecofeminism: Women, Animals Nature</u> ed. Gaard, Greta (Philadelphia: Temple University Press, 1993)

Leopold, Aldo <u>A Sand Couny Almanac and Sketches Here and There</u>
New York: Oxford University Press, 1968, c.1949)

Lepkoff, Daniel, <u>Contact Quarterly</u> Winter/Spring 2000 pg. 62

Locke, John <u>An Essay Concerning Human Understanding</u> in <u>Great Books of the Western World</u> Robert Maynard Hutchins, Ed. (Chicago: Encyclopedia Brittanica, 1952)

Machamer, Peter "Galileo's Machines, His Mathematics, and His Experiments" in <u>The Cambridge Companion To Galileo</u> ed. Peter Machamer (Cambridge: Cambridge University Press, 1998)

McKibben, Bill <u>Enough: Staying Human in an Engineered Age</u> (New York: Times Books, 2003)

McMullin, Ernan "Galileo Galilei" in <u>Routledge Encyclopedia of Philosophy</u> Gen. Ed. Edward Craig (New York: Routledge, 1988)

Merchant, Carolyn <u>Radical Ecology</u> (New York: Routledge, 1992)

Mopsik, Charles "The Body of Engenderment in the Hebrew Bible, the Rabbinic Tradition and the Kabbalah" in <u>Fragments for a History of the Human Body</u> Michael Feher ed. (New York: Zone, 1989)

Naess, Arne 'The Shallow and the Deep, Long-Range Ecolog Movement.' Inquiry 16: (1973) 95-100

Naess, Arne and Sessions, George "Simple in Means, Rich in Ends" i Deep Ecology for the 21st Century Ed. by George Sessions (Boston Shambhala, 1995)

Nietzsche, Friedrich Beyond Good and Evil Trans. by Walte Kauffman (New York: Vintage Books, 1989, c.1966)

Nelson, Lisa and Stark Smith, Nancy "Interview with Bonni Bainbridge Cohen" Contact Quarterly Winter 1980

Orr, David Earth in Mind: On Education, Environment and the Human Prospect (Washington D.C.: Island Press, 1994)

Paul The Writings of St. Paul ed. Wayne A. Meeks (New York: W.W. Norton and Company, 1972)

Paxton, Steve interviewed by Elizabeth Zimmer on CBC Radio Contact Quarterly Fall 1977 Vol. 3 #1 pg. 11

_____, Contact Quarterly Vol. 4 #2 Winter '79 pg. 26

_____, Contact Quarterly Vol. 7 #3/4 Spring/Summer 1982 pg. 17

Pepper, David <u>Modern Environmentalism</u> (London: Routledge, 1996)

Robinson, George <u>Essential Judaism: A Complete Guide to Beliefs, Customs and Rituals</u> (New York: Pocket Books, 2000)

Roose-Evans, James "Circling the Mountain" <u>Passages of The Soul</u> from Personal Papers of Ann Halprin, NYLPA MGRL 98-4751

Rossi, Paolo <u>The Birth of Modern Science</u> Translated by Cynthis de Nardi Ipsen (Blackwell: Oxford University Press, 2000)

Roszak, Theodore <u>The Cult of Information</u> (New York: Pantheon Books, 1986)

Ruether, Rosemary <u>New Woman/ New Earth: Sexist Ideologies and Human Liberation</u> (New York: The Seabury Press, 1975)

Skinner, Joan interviewed by Agis, Gaby in <u>Dance Theater Journal</u>

Skura, Stephanie "Releasing Dance: Interview with Joan Skinner" <u>Contact Quarterly</u> Fall 1990

Simmons, Ian G. <u>Interpreting Nature: Cultural Constructions of the Environment</u> (New York: Routledge, 1993)

Sondak, Eileen "Using Dance as Ritual for Change" <u>Los Angele</u>
<u>Times</u> August 27, 1985

Spretnak, Charlene "Critical and Constructive Contributions o
Ecofeminism" in Peter Tucker and Evelyn Grem Eds. <u>Worldviews an</u>
<u>Ecology</u> (Philadelphia: Bucknell Press, 1993)

Thompson, William Irwin <u>The American Replacement of Nature: The</u>
<u>Everyday Acts and Outrageous Evolution of Economic Life</u> (New York.
Doubleday, 1991)

Thoreau, Henry David <u>Walden: or Life in the Woods</u> (New York: C.N
Potter, c.1970)

Vernant, Jean-Pierre "Dim Body, Dazzling Body" in <u>Fragments for a</u>
<u>History of the Human Body</u> Michael Feher ed. (New York: Zone, 1989)

Williams, Michael A. "Divine Image – Prison of Flesh: Perceptions of
the Body in Ancient Gnosticism" in <u>Fragments for a History of the Human</u>
<u>Body</u> Michael Feher ed. (New York: Zone, 1989)

Witty, Rabbi Abraham B. and Rachel J. Witty <u>Exploring Jewish</u>
<u>Tradition</u> (New York: Doubleday, 2000)

Worster, Donald Ed. <u>The Ends of the Earth: Perspectives on Modern</u>
<u>Environmental History</u> (NY: Cambridge University Press, 1988)

Zientara, Jerry "A Few (Tight) Marbles From My Teacher's Notebook" Contact Quarterly Spring/Summer 1978 Vol. 3 No. 3/4 pg. 10

Zimmerman, Michael A. "Deep Ecology and Ecofeminism: The Emerging Dialogue" in Reweaving the World: The Emergence of Ecofeminism Irene Diamond and Gloria Feman Orenstein Eds. (San Fransisco: Sierra Club Books, 1990)

_____ , Contesting Earth's Future: Radical Ecology and Postmodernity (San Fransicso: University of California Press, 1994)

Made in the USA
San Bernardino, CA
20 December 2017